ELEMENTARY
QUANTUM MECHANICS

THE WYKEHAM SCIENCE SERIES

General Editors:

PROFESSOR SIR NEVILL MOTT, F.R.S.
Emeritus Cavendish Professor of Physics
University of Cambridge

G. R. NOAKES
Formerly Senior Physics Master
Uppingham School

The aim of the Wykeham Science Series is to broaden the outlook of the advanced high school student and introduce the undergraduate to the present state of science as a university study. Each volume seeks to reinforce the link between school and university levels. The principal author is a university professor distinguished in the field assisted by an experienced sixth-form instructor.

ELEMENTARY
QUANTUM MECHANICS

N. F. Mott

Emeritus Cavendish Professor of Physics
University of Cambridge

WYKEHAM PUBLICATIONS (LONDON) LTD
LONDON and WINCHESTER
SPRINGER-VERLAG NEW YORK INC.
1972

Sole Distributors for the Western Hemisphere
SPRINGER–VERLAG NEW YORK INC. NEW YORK

Cover illustration—The picture on the cover is a modern electron diffrac-tion pattern showing the wave nature of an electron beam.   It is obtained with a thin film of platinum about 100 Å thick grown by vapour deposition on the surface of a silver crystal (Courtesy of D. Cherns, Cavendish Laboratory)

ISBN 0–387–91096–4 (Paper)
ISBN 0–387–91101–4 (Cloth)

Library of Congress Catalog Card Number 76–189453

First published 1972 by Wykeham Publications (London) Ltd.

Printed in Great Britain by Taylor & Francis Ltd.
10–14 Macklin Street, London, WC2B 5NF

PREFACE

QUANTUM mechanics is the branch of physics which describes the behaviour of electrons in atoms, in molecules and in solids. For electrons, and also for the other particles of atomic physics, it replaces Newtonian mechanics. Hardly any scientific concept which depends on the behaviour of electrons can be understood without it, whether this be the spectrum of the hydrogen atom, the chemical bond in the water molecule or the action of the transistor. Quantum mechanics was introduced more than forty years ago and, whatever disputes there may be about the philosophical principles involved in its interpretation, its ability to explain a very wide range of natural phenomena is established without any doubt at all. This book is written in the belief that any serious student of physics or of chemistry should obtain as early as he can some idea of what quantum mechanics is about, and that if he does the concepts of atomic and solid state physics and of chemistry will make much more sense to him.

Quantum mechanics uses more advanced mathematical techniques than are necessary for Newtonian mechanics. These techniques, rather than the new physical ideas, are the main obstacles to most students who are introduced to the subject at an early stage in their study of science. In this book the mathematics is kept as simple as possible. It is assumed that the reader knows how to use the differential calculus to solve simple problems, and is familiar with sines and cosines, simple harmonic motion and the elementary properties of waves. For some students with this level of attainment, a difficulty may be the description of a wave in a form which makes use of complex numbers, namely

$$\psi = A \exp \{2\pi i(Kx - \nu t)\},$$

where ψ is the quantity whose variation with x and t shows wave behaviour, and A is called the amplitude. ψ varies between the values $\pm A$, and for an undamped wave it is a constant. K is here the reciprocal of the wavelength (i.e. the 'wave number') and ν the frequency. We do not believe that this can be avoided in an introduction to quantum mechanics; where it is introduced in this book the reason for it is carefully explained. Also the solution in series of the Schrödinger equation for a particle moving in one dimension has been included, in the belief that, whether or not the reader has reached this point in his study of mathematics, it could be useful to him to see how it is done,

v

since the derivation of the Bohr formula for the energy levels of hydrogen depends on it.

The first few chapters contain an outline of atomic physics. These chapters are not in any sense meant as an introduction to the subject; it is assumed that the reader already knows something of the electron, the atomic nucleus and the structure of the atom. The purpose is to summarize the outstanding facts of physics that quantum mechanics has to explain if it is to do its job. In Chapter 10 some examples of the applications of quantum mechanics are described, particularly from the field of solid state physics, including a brief mention of one problem which at the time of writing is unsolved.

This book is written at a time when SI units have been adopted in schools and universities for the teaching of physics, and they are used in this book. The serious student of atomic physics will however find a confusing variety of units in use in the research journals, and will have to familiarize himself with electrostatic and centimetre-gramme-second units to follow the past and much of the present literature. In this book, in common with almost all original work appearing at present, we retain the electron volt as a unit of energy. We also measure lengths both in nanometres (10^{-9} m) and ångströms (1 Å = 10^{-10} m). We make one further concession to electrostatics. In the theory of the atom, from Rutherford and Bohr onwards, the electrostatic force between an electron and a proton, or another electron, is the basic quantity on which all else depends. If the particles are at a distance r metres from one another this force (in newtons) is written $e^2/4\pi\varepsilon_0 r^2$, where e is the charge in coulombs on each particle and ε_0 the permittivity of free space. If e is in electrostatic units, r in centimetres and the force in dynes, this becomes e^2/r^2, the electrostatic unit of charge being chosen so that this shall be so. Where appropriate, the formulae for various physical quantities will be expressed also in this form.

Finally, it is a pleasure to thank the many people, both students and teachers in schools and universities, who have helped me in planning and writing this book.

Cambridge
December 1971   NEVILL MOTT

CONTENTS

CHAPTER 1

quantization of energy

QUANTUM MECHANICS is the branch of mathematical physics which makes it possible to calculate the properties of atoms. Following the discoveries of J. J. Thomson and Rutherford we have known for more than fifty years that an atom consists of a nucleus and a number of electrons; quantum mechanics makes it possible to calculate such quantities as the ionization energy of an atom, which is the energy required to remove an electron from an atom, or the binding energy of a molecule such as H_2, which is the energy required to separate the two atoms of the hydrogen molecule. To describe the behaviour of electrons, quantum mechanics has to be used instead of Newtonian mechanics. Newtonian mechanics is all that is needed for most of the problems of planetary motion, space travel and mechanical engineering; it fails only for bodies moving very fast, when it must be modified in ways which depend on the principle of relativity, and for particles of very low mass such as electrons, when it must be replaced by quantum mechanics.

The laws of quantum mechanics were discovered in the period 1924–1928. Among physicists it was a period of great excitement; many of the unexplained facts of physics and of chemistry known for fifty years or more seemed suddenly to tumble into place and to become part of a comprehensible pattern. Now, forty years later, quantum mechanics is as well established as Newtonian mechanics. It has become part of the language of physics and chemistry and of some branches of technology such as solid state electronics. But this does not mean that quantum mechanics is a finished subject in which there is nothing more to be done. Its methods are constantly being improved and strengthened in order to tackle complicated problems involving very large numbers of electrons in molecules and in solids. Moreover in nuclear physics, and particularly in the branch of science opened up by the big particle accelerators, it is by no means certain that present-day quantum mechanics combined with relativity is good enough, even though the quantum mechanics of the high energy nuclear physicists has evolved a great deal from what it was in the first decade.

The outstanding success of quantum mechanics at the time of its formulation was the explanation it provided of the 'quantization' of the energy of the atom. The hypothesis that the energy of an atom is quantized was first put forward by the Danish physicist Niels Bohr in 1913, though the concept of quantization of some other systems

1

dates from the work of Max Planck at the beginning of this century, as we shall see in Chapter 2. To see what quantization means, let us first consider what was known at that time about the constituents of an atom. After the discovery of the electron by J. J. Thomson in 1900 and the measurement of e/m_e, it was believed that atoms contained electrons. The Zeeman effect (Chapter 5) provided evidence for this. If this were the case, the atoms would contain positive charge as well and one model due to Thomson envisaged the electrons as embedded in the positive charge like currants in a bun. If so, they could vibrate about their mean positions, each electron with its own characteristic frequency, and so emit or absorb radiation. But Rutherford's experiments in 1911 on the scattering of alpha particles by thin sheets of metal showed that the positive charge was all concentrated in a small heavy nucleus, and that most of the volume occupied by the atom was empty. Moreover the experiments enabled an estimate to be made of the charge (denoted by Ze) on the nucleus, and thus of the number (Z) of electrons in the atom. This, and some work by Moseley on the X-ray spectra of the elements (see Chapter 3), enabled the very important hypothesis to be put forward that Z is to be identified with the position of the atom in the periodic table of the elements, being equal to 1 for hydrogen, 2 for helium, 11 for sodium, 79 for gold and so on. This hypothesis is now fully substantiated.

In the light of these developments, an atom was thought to be similar to a solar system, with a number of comparatively light particles, the electrons, taking the place of the planets and revolving round a heavy nucleus. In particular, the hydrogen atom was seen to contain only a single electron moving round its nucleus, which is the proton. If the electron moves in a circular orbit of radius r, the electrostatic force acting on the electron is $e^2/4\pi\varepsilon_0 r^2$. Here e is the charge on the electron in SI units $(1\cdot602 \times 10^{-19} \text{ C})$ and ε_0 the permittivity of a vacuum $(8\cdot85 \times 10^{-12} \text{ J}^{-1} \text{ C}^2 \text{ m}^{-1})$. If the velocity of the electron is v in m s^{-1}, we can equate this electrostatic force to the product of the mass and the radial acceleration v^2/r, obtaining

$$\frac{1}{4\pi\varepsilon_0} \frac{e^2}{r^2} = \frac{m_e v^2}{r}, \tag{1}$$

an equation which gives a relation between r and v. It enables us to write down the energy of the moving electron in terms of r. The kinetic energy is $\frac{1}{2}m_e v^2$. The potential energy, if it is taken to be zero when the electron is at an infinite distance from the proton, is

$$\frac{1}{4\pi\varepsilon_0} \int_r^\infty \left(-\frac{e^2}{r^2} \right) dr = -\frac{1}{4\pi\varepsilon_0} \frac{e^2}{r};$$

the minus within the brackets arises because the force on the electron is *towards* the nucleus, that is in the direction of decreasing r. So the

2

total energy W of the electron, the sum of the potential energy and the kinetic energy $\frac{1}{2}m_e v^2$, is given by

$$W = \tfrac{1}{2}m_e v^2 - \frac{1}{4\pi\varepsilon_0}\frac{e^2}{r}, \tag{2}$$

and substituting for $\frac{1}{2}m_e v^2$ from (1) we find

$$W = -\frac{1}{8\pi\varepsilon_0}\frac{e^2}{r}. \tag{3}$$

So the energy is *negative*; this means that it is lower than it would be if the electron were removed to an infinite distance from the nucleus. $-W$ is the energy required to remove it.

This orbital model of the hydrogen and of more complicated atoms had one difficulty which the 'currant bun' model did not have; an electron moving in an orbit round a nucleus would radiate energy in the form of light; as it lost energy, its remaining energy W would become algebraically smaller, r would become smaller too so that the atom would shrink. What was to stop it? Some mechanism is essential, giving a fixed minimum radius. The idea of a fixed size to an atom is essential both for our understanding of how they fit together in solids and for the kinetic theory of gases. The orbital model did not provide a fixed size.

It was here that Bohr with his hypothesis of the quantization of energy showed the way forward. To show what he meant, we cannot do better than quote from one of his papers (1917). He says that the quantum theory rests upon the following assumption (among others):

"That an atomic system can, and can only, exist permanently in a certain series of states corresponding to a discontinuous series of values for its energy, and that consequently any change of the energy of the system, including emission and absorption of electromagnetic radiation, must take place by a complete transition between two such states. These states will be denoted as the 'stationary states' of the system."

This assumption, that an atomic system can only have a discontinuous series of values for its energies, is what is meant by the quantization of energy.†

Bohr's hypothesis can be divided into two parts.

(*a*) For an atom of a given kind, there is a lowest value W_0 that the energy W can have. For the 'currant bun' model there would be nothing new in this; W_0 would be the energy when the electrons stopped vibrating and $-W_0$ would be the energy needed to get one of them out

† Quantization is sometimes taken to mean that some quantity, such as the angular momentum of an atom or the energy of a monochromatic beam of radiation, is an integral multiple of a natural unit of that quantity, its 'quantum'. In this book we say that a quantity is quantized if it can only have one of a series of discrete values, whether they are a multiple of a natural unit or not.

3

of the atom. But for the orbital atom it was a new and revolutionary hypothesis. Moreover, for the hydrogen atom it showed that the orbit would have a fixed radius a_0, given according to (3) by

$$-\frac{1}{8\pi\varepsilon_0}\frac{e^2}{a_0} = W_0,$$

and thus it gave some idea of how an atom could have a fixed size.

(b) Bohr also assumed that all the higher values of the energy that a system could have would be quantized too; that is to say, again for an atom of a given kind, there would be a series of energy values W_1, W_2 ... which it could have. These are shown in fig. 1. In this figure an arbitrary choice has to be made about the zero of energy. But of course, since we are only concerned with the *changes* in the total energy, our choice of zero does not affect these.

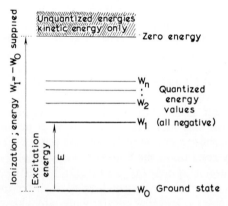

Fig. 1. Energy levels W_0, W_1, W_2 ... of a typical atom. W_0 is the energy of the ground state, W_I the ionization potential and E the excitation potential. Energies in the shaded region of the diagram are unquantized, the electron being separated a large distance from the ion and the energy being the kinetic energy $\frac{1}{2}m_e v^2$ only.

In fig. 1 we choose the energy of an electron at rest at an infinite distance from the atom as zero. This means that all the quantized energies W_n are *negative*; $-W_n$ is the energy which must be *supplied* to remove the electron from the atom. The electron can have *any* positive energy, because, when it is far from the atom, it can move with velocity v_0 say and so have kinetic energy $W = \frac{1}{2}m_e v_0^2$, and this is not quantized. When it moves near to the atom, it will move faster; but the total energy, given by equation (2), remains unquantized. In fig. 1 these unquantized energies are shown shaded.

If an electron has the quantized energy W_n it is said to be in the quantized state n. The state with lowest energy (W_0) is called the ground state.

The energy $(-W_0)$ necessary to remove an electron from an atom is called the ionization energy and will be denoted by W_I, so that

$$W_I = -W_0.$$

The potential by which an electron must be accelerated to give it an energy W_I is written V_I (the ionization potential), so that, if W_I is measured in electron volts,

$$eV_I = W_I,$$

where e is the charge on the electron.

Several points about this model may be emphasized. It applies to free atoms, for instance in a monatomic gas, not to atoms combined in molecules or solids. It is true for all atoms, and for that matter for molecules and for the energies of the 'nucleons' in nuclei. It is by no means restricted to hydrogen. It has survived the introduction of quantum mechanics and indeed been confirmed by it, though the orbital theory (Chapter 3) which Bohr used to calculate the quantities W_n for the simplest atom, hydrogen, has not. Also it has been amply confirmed by experiment. Some of the experimental evidence will now be described.

In the experiments which give the most direct evidence, electrons in a beam in which they have been accelerated through a known potential designated by V_0 are allowed to collide with the atoms of a gas. If the energy $\frac{1}{2}m_e v^2 = eV_0$ of each electron is less than $E = (W_1 - W_0)$, then the electron cannot transfer any energy to the atom and will bounce off it elastically. $(W_1 - W_0)$ is known as the first excitation energy, which we denote by ΔW so that

$$(W_1 - W_0) = E.$$

If eV_0 is greater than E, many of the electrons on hitting an atom will transfer energy equal to E to it, and bounce off with energy

$$eV_0 = E.$$

This behaviour can be observed in several ways, and early experiments designed for this purpose are important for establishing the quantum theory.

One of the most famous of these experiments is that of Franck and Hertz. The apparatus which they used is shown schematically in fig. 2 (a) and in more detail in fig. 2 (b). Electrons were given out by a hot wire, which is marked W, into a vessel containing mercury vapour at low pressure. They are then accelerated towards a gauze cylinder which is marked G. If the accelerating voltage V_A is less than the first excitation energy ΔV given by $E = e \Delta V$ (see fig. 1), then no electrons will suffer inelastic collisions with mercury atoms between the wire and the gauze. However many atoms they hit in

5

their passage between them, they will arrive at the gauze with energy near to eV_A. Between the gauze and an earthed cylindrical electrode marked R in the figures is another smaller potential difference V_R

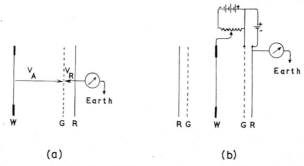

Fig. 2. Apparatus for the Franck–Hertz experiment. (*a*) Shows the scheme, and (*b*) the actual arrangement of the cylindrical electrodes.

which opposes the motion of the electrons. As long as V_R is less than V_A, which means that W is negative with respect to R, the electrons ought to arrive at the earthed electrode and be observed through the current they produce to earth through the galvanometer.

If however the potential V_A is greater than the first excitation potential of the atom concerned, a great many of the emitted electrons will lose energy when they make collisions with the atoms. Many of them therefore will arrive at the gauze G with much lower energies than eV_A. If the reverse potential V_R has an appropriate value (i.e.

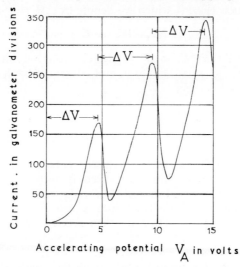

Fig. 3. Current–voltage curve in the Franck–Hertz experiment. The first drop occurs at a voltage ΔV, and successive ones with further increments ΔV.

6

$V_R > V_A - \Delta V$), these electrons will not get to the receiving plate. Therefore, if the potential V_A is increased, a drop in the current through the galvanometer is to be expected when V_A exceeds the first excitation potential. Franck and Hertz carried out the experiment using mercury vapour and their results are shown in fig. 3. A drop in the current occurs not only when V_A rises to the value $\Delta V = E/e$, but again when it reaches twice or three times this value so that the electrons on the way from wire to gauze suffer two, or three, collisions.

In these experiments the pressure of the gas has to be such that each electron suffers several collisions on its way through. Another kind of experiment has been used in which the pressure of the gas is very low, so that most of the electrons in a beam pass through it without any collisions and those which have hit a gas atom and are scattered out of the beam have rarely hit more than one atom. The mechanism of energy loss can therefore be investigated in detail. One expects, following Bohr's hypothesis of stationary states, to find that some of the scattered electrons have lost no energy at all, some have lost the energy $E = (W_1 - W_0)$, others have lost rather more energy and indeed energy equal to any of the discrete values $W_n - W_0$, whereas others will have lost any amount of energy greater than the ionization

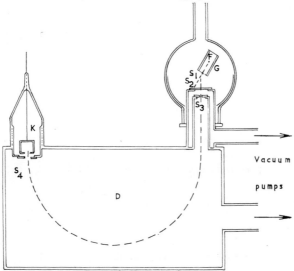

Fig. 4. The Dymond–Watson experiment for electron scattering in helium. The magnetic field in D is at right angles to the plane of the diagram and upward.

energy W_I after hitting an atom and knocking an electron off it. An experiment of this kind carried out in the Cavendish Laboratory by Dymond and Watson† in 1928 will now be described. Their apparatus is shown in fig. 4. Electrons are produced by a hot tungsten filament

† E. G. Dymond and R. Watson, *Proc. R. Soc.* A, **122**, 571, 1929.

at F and are accelerated in the metal box G by a voltage, which can be varied, between the wire and the surface of the box. Only those electrons which arrive at the narrow slit S_1 get out; they can be considered as having all the same energies because the energy due to the accelerating potential difference (say 100 volts) is much greater than their thermal energies (say 0·1 of an electron volt). They then enter a spherical vessel containing helium gas. Helium was chosen because it is monatomic. Since the pressure is such that most electrons will not suffer more than one collision, those which pass through the two further slits S_2 and S_3 ought either to have lost no energy at all as a consequence of a single elastic collision, or else they ought to have lost energy E or one of the other excitation energies; alternatively they will have ionized an atom and lost an energy greater than E.

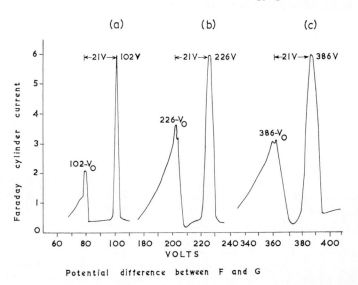

Fig. 5. Current–voltage curves for the Dymond–Watson experiment for three different initial accelerating voltages (a) 102 V, (b) 226 V, (c) 386 V. In each case the peak indicating energy loss appears 21 V below the initial voltage.

For helium, the quantity E is 21 eV and W_I is 24 eV. After passing through the slits the electrons pass through into a chamber D which is kept evacuated by pumps which remove any gas which passes through the slits. A magnetic field is applied across this chamber, so that the electrons are deflected into a circular path of which the radius depends on the field and on their velocity. By varying the field one can therefore arrange that electrons of a given velocity pass through the slit S_4 into the Faraday cylinder K and the current entering the cylinder can be measured. The results, showing that the current is a function of the energy of the electrons after scattering, are shown in fig. 5 for three initial accelerating voltages, 102, 226 and 386 V and for a scattering angle of 10 degrees. In both cases the plot of current against energy shows a sharp peak, at an energy equal to that in the original beam; this shows that many electrons are deflected by the helium atoms and do not lose any energy at all. The most striking feature of the results, however, is that, apart from the experimental background, no electrons lose less than 21 eV. This as we have seen is the excitation energy of the helium atom. The resolution of the experiment is barely sufficient to show excitation to the higher states. If the loss of energy is greater than 24 eV, an electron is knocked out of the atom and its kinetic energy can have any value and is not quantized. It will be seen in fig. 5 that values of the energy loss greater than this show no peaks and thus no signs of quantization, which is what one expects.

These two experiments, and many others like them, give some of the most direct experimental evidence of the existence of stationary states. But the use of electrons as bombarding particles does not give very accurate values of the energies of these states and these are obtained through experiments involving the interaction of atoms with light. These will be discussed in the next chapter.

The excitation potentials of most atoms have a magnitude of several electron volts. In a gas temperature at T (in kelvins) the mean kinetic energy of each molecule is $\frac{3}{2}kT$, where k is the Boltzmann constant defined as $k = R/N_A$, where R is the gas constant and N_A is the Avogadro constant. At room temperature kT is equal to about 0·025 eV (4×10^{-21} J). Therefore, in a monatomic gas, the collisions between the atoms do not have nearly enough energy to excite any of them (or at least, the probability of an exciting collision is so remote that we can neglect it). Indeed, were it not so, the whole kinetic theory of gases would be incomprehensible. The molar heat capacity at constant volume of a monatomic gas is $\frac{3}{2}R = \frac{3}{2}N_A k$. If collisions caused any extra motion within the atom, this would give an additional term in the energy, which would increase with T, and give something extra to the molar heat. Without the quantum theory, the model of the gas atom as a hard elastic sphere, on which the theory of specific heat is based, is quite inexplicable.

The molar heat capacity at constant volume of a gas of diatomic molecules is $\frac{5}{2}N_A k$. The additional contribution $N_A k$ compared with the monatomic gas comes from the mean energy kT of rotation of each molecule, which has two rotational degrees of freedom.† In

† 'Degrees of freedom' means the number of coordinates needed to specify the motion fully; for if one coordinate is unspecified there is freedom for that coordinate to have any value. A point molecule has three degrees of freedom (x, y, z) for its translational motion, and a dumbbell molecule has in addition coordinates θ and ϕ for rotations about independent axes which are at right angles to one another and to its length. Classical statistics indicates that the kinetic energy associated with each degree of freedom is $\frac{1}{2}kT$; so if three degrees of freedom give $\frac{3}{2}N_A k$ for the molar heat capacity, five degrees of freedom give $\frac{5}{2}N_A k$.

hydrogen (H_2), however, the molar heat capacity drops and approaches the value $\frac{3}{2}N_Ak$ at temperatures below that of liquid air. Rotation therefore appears to be stopping. This suggests that the rotational energy of the molecule is quantized, and that the energy required to start if rotating is large compared with the small value kT at low temperatures but small compared with kT at room temperature. The reason why this is so is described in Chapter 10.

properties of radiation

THE quantum theory had its origins in 1900 with the work of Max Planck, who was seeking an explanation of the properties of black-body radiation, that is to say the radiation in thermal equilibrium with hot matter as in a furnace. He introduced the hypothesis that the energy of a light wave is quantized. This hypothesis in a modern form is as follows. A light wave contains energy, and when it falls on the human eye or on a photographic emulsion it gives up energy to it, causing changes such as the decomposition of the silver bromide in an emulsion into silver and bromine. Until Planck's work there was no reason to believe that there could be any restrictions on the value of the energy given up. But following Planck we now believe that the energy of a light wave, just like the energy of an atom, is quantized. By 'quantized' we mean, just as in the discussion of the last chapter, that

(a) The energy of a light wave at a given moment can only have one of a discrete series of values;

(b) unlike the energy of an atom these energies are integral multiples of a particular quantum of energy. This quantum is hv, where v is the frequency of the light and h is the Planck constant, a new constant of Nature equal to $6 \cdot 6 \times 10^{-34}$ J s.

Thus it is believed that the energy carried by monochromatic radiation of frequency v can only have the values nhv, where n is an integer. This means, as first pointed out by Einstein in 1905, that when light gives up its energy to matter, the amount of energy that it can transfer must be a multiple of hv. Actually in giving energy to an electron, or an atom or molecule, a beam of light can normally transfer only *one* quantum at a time. So it transfers an amount of energy equal to hv.

The most striking evidence that light gives up its energy to matter in quantized amounts is provided by the photoelectric effect, which is the ejection of electrons from a metal by the action of light. This phenomenon was discovered in the late nineteenth century, and was explained by Einstein in 1905 in a way that was not generally accepted until after Bohr's work when quantum theory was all the rage and new experiments (notably by R. A. Millikan in 1916) confirmed Einstein's predictions in every respect. The basic apparatus is shown in fig. 6. Light, or ultra-violet light, falls on a metal surface, the emitter (cathode). Any electrons ejected from the metal by the light are accelerated by a voltage between the battery and another electrode, the collector (anode).

A very sensitive galvanometer G registers the rate of transfer of electrons from C to A as a current. The space between the electrodes is evacuated. A high enough voltage must be used to ensure that all electrons emitted move across the evacuated chamber fast enough to avoid setting up a space charge, which would produce a field that would push some of them back to the emitter.

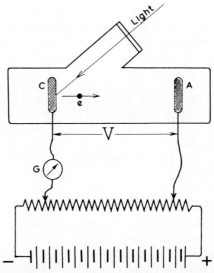

Fig. 6. Production of photoelectrons as in original experiments by Lenard. Light entering the window strikes the cathode C, and there liberates photo-electrons which are accelerated or retarded in the field between C and A.

The unexpected fact about this phenomenon was that, whatever the intensity of the light, *no* electrons are emitted and no current measured if the frequency of the light is below a critical frequency v_0, which depends both on the nature of the metal and on the state of the surface†. Einstein explained this by the following assumptions:

(*a*) It takes an energy equal to or greater than some critical value W_0 to remove an electron from the atom.

(*b*) Light of frequency v can give energy to matter, and thus to the electrons in the metal, only in quanta of amount hv.

(*c*) Therefore no electrons will be emitted if the frequency of the light is less than v_0 where

$$hv_0 = W_0. \tag{1}$$

The frequency v_0 is called the photoelectric threshold and the energy W_0 is called the work function. Values of W_0 range from 4×10^{-19} to 9×10^{-19} joules, that is, from 2 to 5 eV; the corresponding wavelengths,

† Adsorbed oxygen and oxide films frequently increase the value of v_0.

12

given by $\lambda = c/\nu_0$, where c is the velocity of light, range from 500 nm (5000 Å), that of blue light, to far into the ultraviolet.

After 1913, when the quantized nature of the allowed energies of atoms was established, the absorption of light by a gas of monatomic molecules gave an equally direct proof that light has this property. It also confirmed that light normally gives up only *one* quantum at a time. A monatomic gas will therefore absorb light with frequencies ν_n given by

$$hv_n = W_n - W_0, \qquad (2)$$

where W_n are the discrete energies of the excited states and W_0 the energy of the lowest state, known as the ground state, as illustrated in fig. 1. The existence of a complicated line spectrum for each kind of atom, the most interesting feature of which is that the wave-numbers for a given atom could be classified into several series of the form $1/\lambda = A_0 - A_n$, each wave-number being shown as the difference between two 'spectral terms', was one of the facts (together with Planck's hypothesis) which first led Bohr to propose the quantization of energy levels.

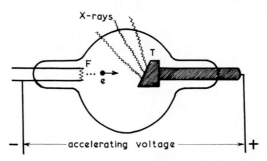

Fig. 7. An X-ray tube. Electrons, emitted from the heated filament, are accelerated to the target from which X-rays are emitted.

We have now to ask whether there is evidence that in the opposite process, in which matter gives up energy to radiation, it does so in quanta of energy $h\nu$. The most direct evidence, the photoelectric effect the other way round so to speak, is provided by the emission of X-rays. An X-ray tube is shown in fig. 7. A beam of electrons emitted from the hot filament F is accelerated through a voltage of at least several kilovolts and falls on a target T (the anticathode). X-rays are emitted from the target. If the methods of X-ray spectroscopy (Chapter 4) are used to separate the emitted radiation into its constituent wavelengths, a plot of intensity against wavelength will appear as in fig. 8. If the voltage is high enough certain sharp peaks (X-ray spectral lines) occur, which depend on the elements in the target and will be discussed in the next chapter. The main background radiation is distributed continuously over all wavelengths up to a minimum wavelength $\lambda = \lambda_0$. This wavelength does not depend on the material of the target used, but does depend on the maximum kinetic energy $\frac{1}{2}m_e v^2$ of each electron, in other words on the voltage at which the tube is working. If we write ν_0 for the frequency of this minimum wavelength, so that $\nu_0 = c/\lambda_0$, the relationship is

$$\tfrac{1}{2}m_e v^2 = h\nu_0. \tag{3}$$

The explanation is that, when hitting an atom, an electron can give up part of its energy to radiation, but has at most only the energy $\frac{1}{2}m_e v^2$ to get rid of (or less if it has been slowed down in penetrating the metal); so no electrons can get rid of an energy greater than $h\nu_0$. Therefore no X-radiation of frequency greater than ν_0 can be produced.

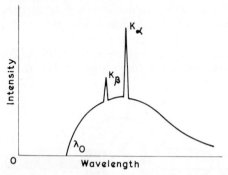

Fig. 8. Intensity distribution with wavelength of X-radiation emitted for voltage about 50 kV. The two peaks K_α and K_β are 'lines' characteristic of the target metal. A limited number of other sharp lines are normally observed if the voltage is high enough to give spectral lines at all.

Equally direct evidence comes from the line emission spectra obtained from free atoms, either in a monatomic gas or in a discharge tube in which a proportion of the molecules are dissociated as a consequence of impact with electrons. As we have seen in the last chapter, when an electron hits an atom, the atom is often left in an excited state. The atom does not stay very long in the excited state; after a time, usually of the order 10^{-8} s, it falls to a lower state giving off energy

$$h\nu = W_n - W_{n'}, \tag{4}$$

where W_n, $W_{n'}$ are the quantized energies of the initial and final states, as illustrated in fig. 1. The frequency of each spectral line emitted is given by an equation of this kind with the appropriate values of W_n, $W_{n'}$. Since the wavelength of light can be measured with very high accuracy, this is far and away the most accurate method of determining the allowed energies W_n of any given atom.

We have to ask, why does an atom in an excited state stay in that state for some time and then decide to give up its energy in the form of light? The discussion of the cause of this behaviour is outside the

14

scope of this book, and here we shall simply remark that, for a given transition between a state n and a state n', there exists a definite probability per unit time (denoted by $A_{nn'}$ and called the Einstein A-coefficient), that the transition occurs. Suppose then that N atoms are left in an excited state, for instance by a pulse of electrons passing through the gas. Then the number that will emit light in a time interval equal to dt is, from our definition of A, equal to $AN\,dt$. The number decreases, therefore, according to the equation

$$\frac{dN}{dt} = -AN$$

which has the solution

$$N = N_0 \exp(-At),$$

where N_0 is the value of N at time zero. The number remaining, and thus the intensity of the light emitted, will therefore decay exponentially.

In the same way, absorption of light can be described by the Einstein B-coefficient, defined in the following way. Suppose that white light, in which there is energy $I(v)dv$ per unit volume in the frequency range v to $v+dv$, acts on an atom in the ground state (energy W_0). Suppose that an excited state exists with energy W_n and

$$h v_{0n} = W_n - W_0.$$

Then the chance per unit time that the atom absorbs a quantum is

$$I(v_{0n})B_{0n}. \tag{5}$$

A method of using quantum mechanics to calculate B_{0n} is described in the last chapter of this book.

2.1 The harmonic oscillator

We have stated that Planck's hypothesis about the quantization of the energy of a light wave was the beginning of the quantum theory. It was Einstein who pointed out that, since a light wave is a system in which the electric field E at any point is varying with time according to the equation

$$E = E_0 \cos 2\pi v t, \tag{6}$$

then any system involving a vibration with frequency v might have this property, namely a quantized energy of amount nhv. One example of such a system is the simple harmonic oscillator. Consider a particle of mass M, bound to a point O such that, when displaced a distance x from a point O, it is pulled back with a force qx. The equation of motion of the particle is

$$M\frac{d^2x}{dt^2} = -qx, \tag{7}$$

15

of which a typical solution is

$$x = x_0 \cos 2\pi \nu t, \tag{8}$$

with

$$\nu = \frac{1}{2\pi} \sqrt{\frac{q}{M}}. \tag{9}$$

Such a particle is said to move with 'simple harmonic motion'. The kinetic energy is $\frac{1}{2}Mv^2$, the potential energy (which is the work done in taking the particle from O and giving it a displacement x) $\frac{1}{2}qx^2$, and the total energy

$$W = \frac{1}{2}Mv^2 + \frac{1}{2}qx^2. \tag{10}$$

It was suggested by Einstein that W should be quantized and only the values equal to $nh\nu$ allowed. The treatment based on wave-mechanics given in Chapter 7 shows that this is very nearly correct, but the formula should be replaced by $W = (n+\frac{1}{2})h\nu$, where n is any integer including zero.

The problem to which Einstein applied this hypothesis is that of the specific heat of a solid. In a solid each atom is vibrating about a mean position, and in a simplified model it is possible to think of a restoring force qx for a displacement of an atom equal to x, so that each atom vibrates with the frequency ν given by (9). Now, as we have stated in Chapter 1, the mean value of the kinetic energy of a molecule in a gas is $\frac{3}{2}kT$. If in a solid there were no quantization of the energy, the kinetic energy of an atom in a solid would have this value too, and it can be shown that the mean value of the potential energy is equal to the same amount. The mean value of the total energy W is thus $3kT$, and the molar heat capacity of a (solid) element is thus equal to $3N_A k$, where N_A is the Avogadro constant, and thus to $3R$, where R is the gas constant, and is the same for all elements in the solid state. This is approximately so at sufficiently high temperatures and is known as the law of Dulong and Petit.

But it is well known, also, that the specific heats of all solids drop progressively towards zero as the temperature is lowered. The explanation given by Einstein is that each vibrating atom has either no energy, or energy equal to $h\nu$, $2h\nu$, and so on. If the temperature is such that

$$kT \ll h\nu,$$

very few atoms will have any energy at all. (The fraction that do is given by the 'Boltzmann factor' $\exp(-h\nu/kT)$, as is shown in books on kinetic theory.) The energy tends to zero much more rapidly than $3RT$, and the molar heat capacity drops to zero too. An example of this behaviour is shown in fig. 9.

16

It is important to understand why the energy $h\nu$ is of order (say) 0·02 eV and thus comparable with thermal energies, in contrast with the excitation energies of atoms, which are of order 5–20 eV and much greater than kT. The forces acting on atoms in a crystal, and the forces acting on electrons in atoms, are all of electrostatic origin and depend on the attraction or repulsion between electrons and ions. We should not expect the difference to lie here. On the other hand, the mass M of an ion is very much greater than the mass m_e of an electron, and according to equation (9), for given restoring force, the mass comes into the formula for the frequency as $1/\sqrt{M}$. Thus we might then expect that $h\nu$ for a vibrating atom should be smaller by a factor $\sqrt{(m_e/M)}$ than a typical excitation energy of an atom.

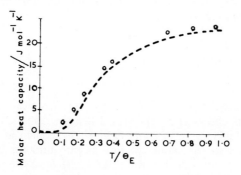

Fig. 9. Comparison of the experimental values of the molar heat capacity of diamond (circles) and values calculated from Einstein's model (Einstein, *Annalen der Physik*, **22**, 180, 1970). The temperature Θ_E is defined as $h\nu/k$ and the fit was obtained with $\Theta_E = 1320$ K (from Kittel, *Solid State Physics*, p. 124).

Let us see how this looks for lead. The relative atomic mass of lead is 207·2 and the mass of a hydrogen atom is $1840 \times m_e$ so for lead M/m_e is about 381 000. The factor $\sqrt{(m_e/M)}$ is thus about 0·0017. Figure 9 shows that the molar heat capacity drops off in the neighbourhood of the temperature of liquid air, so $h\nu$ should be of order $1/100$ eV. So, since $kT \simeq 1/40$ eV at room temperature, these very crude considerations explain the order of magnitude of the difference quite well.

2.2 The light quantum or photon

The photoelectric effect is often described by saying that a beam of light, when it exchanges energy with matter, behaves like a beam of particles each with energy $h\nu$ and moving with velocity c. Whether one describes the phenomenon this way, or whether one says that a light wave is an electric field moving at all points with simple harmonic motion and *therefore* with quantized energy $n h\nu$ (i.e. with n quanta each equal to $h\nu$), is largely a matter of taste; in more advanced discussions the two treatments do not appear different. But the particle

17

model is certainly very valuable for visualizing some phenomena, notably the Compton effect, which will now be described.

The 'particles' are called 'light quanta' or 'photons'. They can of course be created or destroyed when matter and light exchange energy; they are not conserved in the way electrons normally are.† An important quantity is the momentum of a photon. Since a photon moves with the velocity of light, its momentum can only be deduced by using formulae drawn from the theory of special relativity. The relation between the energy W of a particle in this theory and its momentum p is

$$W = c\sqrt{(m^2c^2 + p^2)}, \tag{11}$$

where m is the mass when the particle is at rest and c is the velocity of light. A photon disappears when it is not moving; it does not have any mass so m must be put equal to zero. Thus for a photon

$$W = pc \tag{12}$$

and the momentum p, since $W = h\nu$, is given by

$$p = h\nu/c = h/\lambda. \tag{13}$$

The Compton effect, discovered by the American scientist A. H. Compton, was observed by allowing a beam of monochromatic X-rays to fall on a thin sample of some light element such as carbon. (The outer electrons in the atoms of low atomic number behave for this purpose as if they were completely free.) Most of the radiation passed

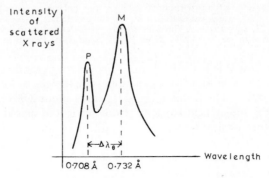

Fig. 10. The Compton effect; wavelengths of quantum elastically and inelastically from an atom.

through the specimen without effect, but some was scattered. Compton studied the wavelength distribution of radiation scattered through some definite angle, say 90°. He found, plotting intensity of the radiation against wavelength, two peaks, P and M (fig. 10). For P, this is the original radiation scattered without change of wavelength; for M, the

† Electrons and positrons can be created by high frequency radiation; the charge, not the number of particles, is conserved.

18

wavelength of the scattered radiation has increased, and $\Delta\lambda_\theta$, the increase in the wavelength, depends on the scattering angle.

The observation is obviously similar to the Dymond–Watson experiment described in Chapter 1, in which an electron is scattered by an atom elastically or inelastically. But in this experiment the energy $h\nu$ of the X-ray photon is much greater than the binding energy of the electron and so one can treat the electron as if it were free. A collision takes place between the electron and the photon in which the photon loses *part* of its energy and in which momentum is conserved. The

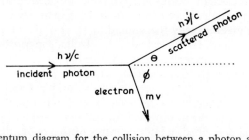

Fig. 11. Momentum diagram for the collision between a photon and an electron (virtually free) in an atom in the Compton effect. The momentum of the incident photon is $h\nu/c$, that of the scattered photon $h\nu'/c$, and that of the ejected electron $m\nu$.

behaviour of both particles is shown in fig. 11. After the collision the photon has a lower energy $h\nu'$ than before, so radiation is observed as having a longer wavelength. The conservation of energy gives

$$h\nu = h\nu' + \tfrac{1}{2}m_e v^2,$$

and the conservation of momentum

$$\frac{h\nu}{c} = \frac{h\nu'}{c}\cos\theta + m_e v \cos\phi$$

and

$$\frac{h\nu'}{c}\sin\theta = m_e v \sin\phi.$$

The elimination of ϕ and v from these equations leads to the equation for the Compton shift

$$\Delta\lambda_\theta = \frac{h}{m_e c}(1 - \cos\theta),$$

valid if $\Delta\lambda_\theta/\lambda$ is small. This equation has been found to agree well with experiment, and the recoil velocities of the ejected electrons have been measured too.

We see then, that when a free electron scatters light it receives momentum from the light. This fact is of importance for the discussion of Heisenberg's uncertainty principle (Chapter 11).

19

CHAPTER 3
the old quantum theory

THE preceding chapters of this book have given evidence, based on collisions with electron beams and emission and absorption of radiation, that the energy of the electrons in an atom is quantized, but no indication has been given of why this should be so or of how the values which the energy may have can be calculated. In his papers of 1913, in which he postulated the quantization of the energy of an atom, Niels Bohr put forward also a hypothesis from which he calculated the values of the quantized energy of the electron in the simplest atom, namely that of hydrogen. Though this hypothesis and the calculations that go with it have now been replaced by others derived from quantum mechanics and described in later chapters of this book, they are of great historical interest and so near to what we think now that it is well worth understanding them.

We must now refer briefly to the great accumulation of very accurate wavelength determinations by spectroscopists, which were already there to be used. Since each element gave its own characteristic pattern of spectral lines, this would quite obviously contain information about the structure of its atoms if only the clue could be found. The classification of the wavelengths of the lines in the spectrum of an element into series, of which a typical term could be written down, was a highly developed art during the later part of the nineteenth century. This developed even further when it was found that the wave-number for members of a given series could be expressed as a difference between two terms, of which one was fixed and the other involved a sequence of integers. The letters s, p, d, f now used to denote azimuthal quantum numbers 1, 2, 3, 4 originally stood for 'sharp', 'principal', 'diffuse' and 'fundamental' series. And in the *spectral terms* we can see that the spectroscopists had clearly identified energy levels. But the patterns were complicated and the wealth of data to be handled tremendous; one wonders how they ever managed at all without a model to give the data significance.

The simplest and most remarkable of these series is the Balmer Series for four lines in the spectrum of atomic hydrogen. The wavelengths in ångströms are: H_α, 6562, H_β, 4861, H_γ, 4340 and H_δ, 4102. Converting them into metres, the wave-numbers form a series

$$1/\lambda = R_H(1/2^2 - 1/n^2), \quad \text{where } n = 3, 4, 5, 6$$

going from α to δ. R_H is the Rydberg constant *for hydrogen*. Its value to four significant figures (though even then it was given to six) is $1 \cdot 097 \times 10^7$ m^{-1}. Since, as we now understand, it plays a fundamental part in the calculation of quantized energy levels, the Rydberg constant itself (now established with certainty to seven significant figures) can be expressed in terms of the fundamental constants including e, m_e and h.

The value of R_H works out to be $m_e e^4/8\varepsilon_0^2 h^3 c$, according to quantum mechanics. Substitution of the numerical values shows at once how closely this particular problem has been solved. It was, indeed, the first problem to be solved by what we now call the old quantum theory, by Niels Bohr in 1913.

In 1913 it was only two years since Rutherford had established the nuclear model of the atom, and so it was known that the hydrogen atom consisted of a single electron together with a proton. The attractive force between the two was thought to be the electrostatic attraction between two point charges of opposite sign and thus equal to

$$\frac{1}{4\pi\varepsilon_0} \frac{e^2}{r^2}$$

when the particles are at a distance r from one another. So the electron should be able to go round the much heavier proton in elliptic or circular orbits. For circular orbits the Coulomb electrostatic force should be equal to the mass multiplied by the acceleration towards the nucleus, so that

$$\frac{1}{4\pi\varepsilon_0} \frac{e^2}{r^2} = \frac{m_e v^2}{r}, \tag{1}$$

as we have already seen in Chapter 1. This equation relates the velocity v of an electron in a circular orbit to its radius. Any value of r would be allowed according to Newtonian mechanics. The total energy W is the sum of the kinetic energy and the potential energy $V(r)$; we define the potential energy $V(r)$ as before as the work required to bring up the electron from infinity, which is given by

$$V(r) = \int_r^\infty \left(-\frac{1}{4\pi\varepsilon_0} \frac{e^2}{r^2} \right) dr = -\frac{1}{4\pi\varepsilon_0} \frac{e^2}{r},$$

so that the total energy W is given by

$$W = \tfrac{1}{2} m_e v^2 - \frac{1}{4\pi\varepsilon_0} \frac{e^2}{r}.$$

Substituting for v from (1), the energy becomes

$$W = -\frac{1}{8\pi\varepsilon_0} \frac{e^2}{r}. \tag{2}$$

We have discussed in Chapter 1 the meaning of a negative sign for the energy.

21

If W is to be quantized, that is restricted to a fixed series of values, some restriction has to be placed on either r, v or on some function of them. Bohr chose to restrict the angular momentum, which we write p_θ; this is for circular orbits equal to $m_e v r$. He made this choice because p_θ is a constant of motion for a particle moving in an elliptic orbit as well as in a circular orbit; r and v are not. By quantizing p_θ, therefore, he did not change the orbits allowed by Newtonian mechanics, he only restricted the ones that could occur. Bohr then assumed that the angular momentum p_θ must be restricted to the values

$$p_\theta = \frac{nh}{2\pi}, \quad n = 1, 2, \ldots, \tag{3}$$

where h is Planck's constant.

We next ask whether there is any logical reason behind the introduction of the 2π in equation (3). In Chapter 5 we shall see how (3) can be deduced from quantum mechanics. However, what was already known in 1913 about Planck's constant h made the 2π necessary. Consider a particle attracted to a fixed point not by a Coulomb force but by a 'simple harmonic' force which is proportional to the distance r between the particle and the fixed point. This force may, as in Chapter 2, be denoted by $-qr$. The particle can execute simple harmonic motion with frequency v given by

$$v = \frac{1}{2\pi} \sqrt{\left(\frac{q}{m_e}\right)}. \tag{4}$$

It was known at the time of Bohr's work that the energy of a vibrating particle must be quantized and equal to nhv, where n is an integer as described in the last chapter. But such a particle can also go round in a circle, with as before the attractive force producing the acceleration towards the centre, so that for the electron

$$m_e v^2 / r = qr. \tag{5}$$

The total energy of the electron is

$$\tfrac{1}{2} m_e v^2 + \tfrac{1}{2} q r^2$$

which from (5) is equal to $m_e v^2$; this has to be equal to nhv, so

$$m_e v^2 = nhv.$$

The angular momentum p_θ is mvr as before, and eliminating r by using equation (5) this gives

$$p_\theta = m_e v^2 \sqrt{\frac{m_e}{q}} = nhv \sqrt{\frac{m_e}{q}}$$

which by (4) reduces to

$$p_\theta = nh/2\pi.$$

22

So, if Bohr had not introduced the factor 2π, a wrong result would have been obtained for a particle attracted by the force qr and moving in a circular orbit.†

Turning again to the hydrogen atom, the three equations (1), (2) and (3) enable us to obtain values of the energy W and of the radius r of an electron moving in a circular orbit. Thus (3) gives $m_e v r = nh/2\pi$, and eliminating v by means of (1) we find $r = n^2 a_0$, where

$$a_0 = \varepsilon_0 h^2/\pi m_e e^2.$$

The quantity a_0 is called the Bohr radius and is equal to 0.053 nm (0.53 Å). Equation (2) gives for the energy

$$W = -W_H/n^2, \tag{7}$$

where W_H is the ionization energy of hydrogen, which is given by

$$W_H = \frac{m_e e^4}{8\varepsilon_0^2 h^2} \tag{8}$$

and is equal to 2.14×10^{-18} J (13.4 eV).

In electrostatic units a_0 and W_H take the form

$$a_0 = h^2/4\pi^2 m_e e^2 = \hbar^2/m_e e^2,$$

$$W_H = 2\pi^2 e^4 m_e/h^2 = e^4 m_e/2\hbar^2.$$

In the ground state, namely the stable state of the hydrogen atom with the lowest of the possible energies, the energy is $-W_H$.

Equation (7) is in excellent agreement with experiment, the frequencies of the spectral lines emitted by (atomic) hydrogen in a discharge tube being given by the equation

$$h\nu = W_H \left(\frac{1}{n^2} - \frac{1}{n'^2}\right). \tag{9}$$

As $\nu = c/\lambda$, the Rydberg constant R_H of the Balmer series is $W_H/hc = m_e e^4/8\varepsilon_0^2 c h^3$.

It is a very curious fact that quantum mechanics gives the same formula as equation (7) for the energy levels of hydrogen, as we shall see in Chapter 8; the old and new quantum theories agree. This is not so for any form of the potential energy $V(r)$ other than the Coulomb form.

The more recent replacement of Bohr's model by a treatment based on quantum mechanics shows that electrons do not rotate round the atom in orbits; also Bohr's theory is in serious disagreement with quantum mechanics in predicting that the ground state of hydrogen has one quantum of angular momentum; according to quantum mechanics it has none. But the model is extremely useful in giving a qualitative description of the behaviour of atoms other than hydrogen

† The symbol $\hbar = h/2\pi$ is often used to simplify the notation.

23

which contain more than one electron. In particular, it gives a good account of the X-ray spectra and the way these spectra are related to the position of the atom in the Periodic Table. This will now be described.

3.1 X-ray spectra

Rutherford's experiments on the scattering of alpha particles made possible an approximate determination of Ze, the charge on the nucleus, because the theory of the scattering process shows that the number of particles scattered depends, among other factors which could be determined, on Z^2e^2, the square of the charge on the nucleus. In the years immediately following Rutherford's experiments it was realized that Z, also equal to the number of electrons in the atom, could be identified with the position of the element in the Periodic Table. The evidence came mainly from the examination of X-ray spectra, particularly in the hands of Moseley, working with Rutherford in Manchester. The description, based on Bohr's theory, of an atom containing many electrons and the explanation of X-ray spectra that it provides will now be described.

Let us then consider first a single electron in the field, not of a hydrogen nucleus with charge e but of a nucleus of charge Ze. Everywhere in the derivation of Bohr's formulae e^2 has to be replaced by Ze^2. Thus the values of the quantized energies of an electron in the field of the nucleus become instead of (7), since by (8) W_H contains the factor e^4,

$$W_n = -Z^2 W_H/n^2. \tag{10}$$

As before, the negative sign means that the electron moving round the nucleus has a lower energy than an electron at rest in free space. The radius of an orbit is

$$n^2 a_0/Z. \tag{11}$$

For example, the atomic number Z of lead is 82. The energy of the ground state is thus 82^2 times the ground state of hydrogen, and its energy, which is negative, is $82^2 \times W_H = 1\cdot6 \times 10^{-14}$ J which is about 100 000 eV. The frequency ν of a photon of energy $1\cdot6 \times 10^{-4}$ J is $2\cdot5 \times 10^{19}$ Hz, corresponding to a wavelength about $1\cdot2 \times 10^{-11}$ m $(0\cdot12$ Å$)$, which lies well in the X-ray region.

We now consider the atom with several electrons. We ascribe to each electron an orbit with a given quantum number n and a quantized energy. In doing this, we are making an approximation. All our previous discussions have been concerned with the energy of *all* the electrons in the atom. In an exact treatment it is this that is quantized, and it is this that is measured by experiments such as those of Franck and Hertz described in Chapter 1. The electrons in an atom repel one another because each one carries a charge, and to give each of them an orbit and a definite constant energy cannot be quite right. It is not

24

exactly true that each electron has a constant energy, still less a quantized energy. However, the attraction of the nucleus in an atom with a fairly large value of Z is a good deal stronger than the repulsion between any pair of electrons, and so the orbits should not be too much upset by the repulsion.

To understand why this is so and to obtain a model to describe the structure of the many-electron atom, we must make use of the Exclusion Principle. This states that in an atom there can never be more than one electron in a state with given quantum numbers. This very important principle was derived empirically by Pauli in 1925 from line spectra, X-ray spectra and the Periodic Table. Wave mechanics goes some way to giving an explanation of the principle, which will be described in Chapter 9.

We have stated that not more than one electron can be in a state with given quantum numbers. So far we have introduced only a single quantum number, the principal quantum number n which for hydrogen determines the energy of the electron. This is the quantity n which appears in equation (9). The developments of quantum theory in the twelve years following Bohr's papers of 1913 ascribed a number of other quantum numbers to each state, depending on whether the corresponding orbit was circular or elliptical, the orientation of the orbit and so on. These developments are now of only historical interest, since they are replaced by a treatment based on quantum mechanics described in Chapter 9. It will be sufficient in this chapter to state that there are $2n^2$ states with principal quantum number n. Thus for the first few values of n the number of states is as follows:

Principal quantum number	Number of states
1	2
2	8
3	18

In the light of Pauli's principle, then, consider an atom such as copper with atomic number $Z = 29$. Only two electrons can go into the state with $n = 1$, and they have energies $-W_K$, where

$$W_K = 29^2 W_H = 29^2 \times 2.14 \times 10^{-18} \text{ J}$$

which is about 12 000 eV. These two electrons are said to form the 'K shell'† of electrons. An electron accelerated through at least 12 kilovolts would be needed to remove one of these from the atom. The radius of the K shell, given by equation (9), is $0.54/29 = 0.02$ Å, much smaller than the radius of the atom. Up to eight electrons can go into states with $n = 2$ and in copper all these states will be occupied; they form what is called the L shell. Since the two K electrons are

† The sequence of letters K, L, M ... for the Bohr-model 'shells' with principal quantum number $n = 1$, 2, 3 ... is long established. A different system of nomenclature is sometimes used for the *energy levels* themselves.

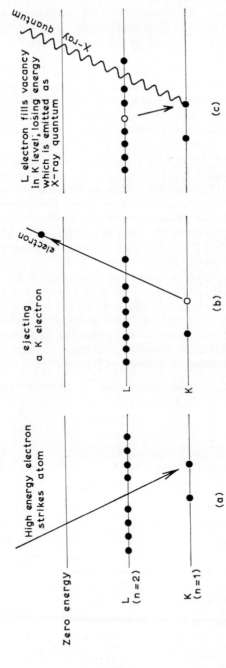

Fig. 12. Mechanism for emission of a characteristic X-ray line (Kα) in an X-ray tube. (a) A fast electron striking a target atom, (b) ejecting an electron, (c) an L electron falling down to this level, with the emission of an X-ray. The continuous background illustrated in fig. 8 does not arise through this mechanism.

much nearer to the nucleus, they reduce the field acting on the L electrons; the average field that pulls them to the nucleus is roughly that of a charge $(Z-2)e$; so their energy is $-W_\mathrm{L}$ where

$$W_\mathrm{L} = 2 7^2 W_\mathrm{H}/2^2$$

which is about 2800 eV. The radius of their orbits is about 0·1 Å. The next 18 electrons go into the M shell, with $n = 3$, and these 28 electrons $(2+8+18)$ together with the nucleus $(Z = 29)$ form the copper ion Cu^+.

There is one electron left over, which in a free atom must have principal quantum number 4. This electron comes off very easily, for instance in copper (I) salts. In metallic copper it is free to move about in the solid and contribute to a current (Chapter 10). Its energy levels will be very dependent on the chemical state of the atom, and in the free atom will have no resemblance to their values in molecules. But this is not so for the inner levels, particularly the K and L levels. They are almost unaffected by the chemical state of the atom. They are responsible for the characteristic lines in the X-ray spectra of atoms (cf. fig. 8), and these lines depend very little indeed on the chemical state of the atom in the bombarded target responsible for the emission of the line.

The mechanism of emission of an X-ray quantum is illustrated in fig. 12. In an X-ray tube (fig. 7) the atoms in the target are bombarded by fast electrons and now and then an electron in a K level is thrown out of the atom. This makes it possible for (say) an L electron to drop into the vacant place left by the ejected K electron, emitting a quantum of X-rays. The frequency ν of the quantum is given by

$$h\nu = W_\mathrm{H} \left\{ \frac{Z^2}{1^2} - \frac{(Z-2)^2}{2^2} \right\}, \tag{10}$$

at any rate approximately. The approximation lies in the neglect of the repulsion between the electrons, which has a small effect on the energies particularly of the L level. Also a more exact treatment (see Chapter 9) shows that there is more than one L level. The line with frequency given by (10) is called the K_α line.

The achievement of Moseley, using the technique of X-ray diffraction discovered by the Braggs, was to discover that the frequency of this line varied in a regular way as one goes from element to element up the Periodic Table. This was just before Bohr put forward his theory. Something, Moseley saw, was changing steadily from atom to atom. This he believed could only be the number of electrons there.

CHAPTER 4
the wave properties of matter

THE old Quantum Theory held the field for the dozen years from its introduction by Bohr until the formulation of quantum mechanics between 1924 and 1926. This was due to the work of theorists such as de Broglie, Schrödinger, Born, Heisenberg, Dirac and many others. The first three developed the concept that particles behave in some ways like waves. Of the greatest importance also for the development of quantum mechanics were the experiments of Davisson and Germer in the United States and of G. P. Thomson in Aberdeen, who in 1926 were the first to show that beams of electrons did indeed have wave-like properties. These experiments, more than any others, made it possible to build up the mathematical framework of quantum mechanics, or wave mechanics as it is sometimes called, on a basis of observed facts. This chapter describes these experiments and their immediate consequences for our thinking about the electron.

It was J. J. Thomson (G. P. Thomson's father) who had established before 1900 that cathode rays consisted of negatively charged particles each with a mass (m_e) very much smaller than the mass of an atom. He established the ratio of charge to mass (e/m_e) by observing the deflection of cathode rays in electric and magnetic fields. Since J. J. Thomson's time much stronger evidence for the particulate nature of electrons has been obtained. Perhaps the most striking is the observation of the tracks made by fast electrons in an expansion chamber. Also atomic theory, as described in the last chapter, equates the position of an element in the Periodic Table (the atomic number Z) to the number of electrons in the atom, an equation which can only be meaningful if electrons are particles. The number of electrons falling on a counter, or the number of electrons in an atom, are quantities that can be measured; in saying that electrons behave in some ways like waves, we must always remember that they *are* particles, in the sense that we can count how many of them there are in a given situation.

In the experiments of Davisson and Germer, a beam of electrons which had been accelerated through a potential of some 100 volts was reflected from the surface of a nickel crystal; the experimental arrangements are shown in fig. 13. A beam of electrons was produced by a hot filament F and accelerated to a metal cylinder A with holes in its walls by a variable voltage V. The beam of electrons struck the nickel crystal at an angle θ_1, and those reflected at an angle θ_2 were collected at C, and the current thus produced was measured by the galvanometer G.

The curious thing about this experiment is that Davisson and Germer were not expecting to find anything peculiar, and only made their famous observation when the nickel crystal had been heated in a way which removed an oxide layer from the surface, so that electrons were reflected from the nickel and not from the oxide which may well have been non-crystalline. What they then found was that electrons were reflected only when $\theta_1 = \theta_2$, and then only for certain angles of incidence. The beam, in fact, behaved like an X-ray beam. The way an X-ray beam is reflected from a crystal will now be described, so that we may then discuss the similar behaviour of electrons.

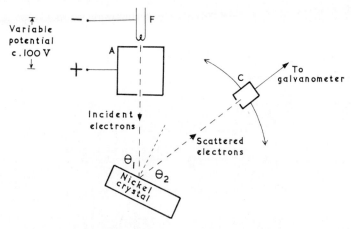

Fig. 13. Experimental arrangement of Davisson and Germer.

X-rays are electromagnetic waves of the same nature as light waves, except that their wavelengths are considerably shorter. The wavelength of blue light is about 400 nm (4000 Å); X-rays can have wavelengths over a considerable range, but typically less than 1 Å, and thus less than the distance between atoms in a crystal. X-rays are emitted when a target (usually metal) is bombarded by electrons with energies usually of several thousand electron volts, and, as explained in Chapter 2, one or more spectral lines of definite wavelength are superimposed on a continuous background.

The reflection of X-rays by crystal planes was first studied by W. H. Bragg and his son W. L. Bragg in 1913. They formulated the 'Bragg law' for the condition under which a crystal surface reflects X-rays, which can be derived as follows. They supposed that every plane of atoms, such as A_nB_n in fig. 14, reflects a small part of the incident wave, although most of it is transmitted. Strong reflection from the crystal will occur, therefore, if the waves reflected from each successive plane are all in phase with one another. This will be so if the path difference between waves from two planes is equal to $n\lambda$, where n is an integer and λ

29

the wavelength. The path difference, using the notation shown in fig. 14, is PQ–PR. If d is the distance between the planes,

$$PQ = d/\sin \theta,$$

$$PR = PQ \cos 2\theta = d \cos 2\theta/\sin \theta.$$

The path difference PQ–PR is thus

$$d(1 - \cos 2\theta)/\sin \theta = 2d \sin^2 \theta/\sin \theta$$

$$= 2d \sin \theta.$$

Thus the waves reflected from all planes of atoms will reinforce one another, giving strong reflection only if the angle of incidence is such that

$$2d \sin \theta = n\lambda. \tag{1}$$

This is the Bragg law.

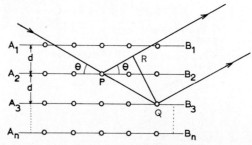

Fig. 14. (a) Arrangement of atoms in a crystal. (b) Illustrating the Bragg law for reflection.

The discovery of Davisson and Germer was that the Bragg law is the condition also for the reflection of a beam of electrons from a crystal surface if one assumes that the beam is behaving like a beam of waves with wavelength

$$\lambda = h/p, \tag{2}$$

where p is the *momentum* of each electron; $p^2/2m_e$ is the corresponding kinetic energy since kinetic energy is $\frac{1}{2}m_e v^2$ and $p = m_e v$. These wavelengths are small; if the electron beam is obtained by accelerating electrons through a voltage V_0, then

$$p^2/2m_e = eV_0.$$

Putting in values of the constants, one finds the following values for the wavelength:

$$V_0 = 10 \quad 100 \quad 1000 \quad 10^4 \quad \text{volts}$$
$$\lambda = 0{\cdot}4 \quad 0{\cdot}13 \quad 0{\cdot}04 \quad 0{\cdot}013 \quad \text{nm}$$

30

Thus, unless rather slow electrons are used, accelerated through less than about 100 volts, the wavelength for electrons is a good deal smaller than the spacings between atoms which lie in the range 0·2–0·4 nm (2–4 Å); the Bragg angles θ given by (1) are therefore small. Modern electron microscopes and diffraction equipment usually work at about 20 000 volts.

Electron diffraction, interpreted according to the Bragg law and using (2) for the wavelength, has developed into an alternative to the use of X-rays for investigating crystal structures. In general electrons do not penetrate as far into solid materials as do X-rays, and electron diffraction is therefore particularly suitable for determining the crystal structure of surface layers, for instance of metal oxides and other products of corrosion on metals. Beams of slow neutrons are also used for certain problems in crystal analysis; equation (2) is true for *any* kind of particle, but wave properties are difficult to observe for any but the lightest particles. If equation (2) is written in the form

$$\lambda = h/\sqrt{(2mW)}, \tag{3}$$

where W is the kinetic energy of each particle, then it will be seen that for charged particles which gain their energy by being accelerated through a few electron volts, λ becomes very much smaller than the distance between atoms in a crystal if m is the mass of any heavy ion. In experiments on neutron diffraction, use is made of neutrons from a reactor which after many collisions with other atoms have been slowed down to have the same thermal energies as the molecules of a gas. The wavelengths for a particle with thermal energies at room temperature are as follows:

neutron (or proton)	H_2	O_2	
10	7	2	nm

while the distance between atoms in a solid is typically 0·3 nm.

Equation (2) (or (3)) looks at first sight rather arbitrary. One wonders why Nature should have chosen just this relationship between an electron's velocity and the properties of a wave. But when the experiments were carried out in 1926, the result was not altogether unexpected. In the first place, the *same* relationship between momentum and wavelength was already known for photons (equation (13) of Chapter 2). Also, the French scientist de Broglie had given a theoretical argument to suggest that *if* there was to be a relationship between a beam of particles and a wave, it must be of this form. De Broglie's full argument, which involves a consideration of vectors and of the Principle of Relativity, will not be described in this book. But its elements are as follows. The momentum p of a moving particle is a vector. If there is a relationship between the momentum of a particle and some property of a wave, then p must be proportional to some

31

vector which describes the wave. A wave travelling in the direction defined by the direction cosines l_1, l_2, l_3 can be written

$$\sin 2\pi\{K(l_1 x + l_2 y + l_3 z) - vt\}.$$

Here K is the wave number, defined as the reciprocal of the wavelength λ, so that

$$K = 1/\lambda.$$

Thus a vector (K_1, K_2, K_3) defines the direction and wavelength of the wave, so the wave number must itself be thought of as a vector, the three quantities K_1, K_2, K_3 being its components along the x, y, z axes, so that $K_1 = Kl_1$, etc. Therefore, de Broglie argued that a relationship of the form

$$K = \text{const. } p$$

was the only one that could exist between the momentum p of a particle and the properties of a wave. Any other relationship would not be invariant under a rotation of axes.

De Broglie also suggested that the frequency v of the wave was related to the energy W of the particle by the relationship

$$W = hv.$$

His proof, based on the Principle of Relativity, will not be given here; an alternative proof is given in Chapter 11.

The experiments on electron diffraction showed then that a beam of electrons, when it passes through a crystal, has to be treated like a beam of waves. If one wants to know where the beam goes, one has to forget that the beam contains particles, and treat it as a wave. Particles will turn up, as a current, as tracks in an expansion chamber or as a darkening of a photographic plate wherever the wave beam is. This is simply what the experiments tell us. It is a natural generalization to assume that *all* predictions of the behaviour of a beam of electrons must be made using a wave model. This assumption is used in many treatments as the basic fact on which quantum mechanics must be built up, and will be so used here. But, starting from the experiments on electron diffraction, one knows nothing *a priori* about these waves. Apart from the wavelength, all we know is that the electrons (or other particles) turn up where the wave is. This statement can be made more precise. We introduce a symbol ψ to denote the *amplitude* of the wave, and define it so that† ψ^2 is the average number of electrons per unit volume in the wave. Or expressed differently, if we take a small unit of volume $dx\, dy\, dz$,

$$\psi^2\, dx\, dy\, dz \tag{4}$$

---

† As we shall see below, ψ is actually a *complex* quantity and we shall have to write $|\psi|^2$ instead of ψ^2.

is the *probability* that an electron will be in the volume element $dx\,dy\,dz$ at any moment of time. From this we conclude that, if the beam of electrons each of which have velocity v is falling normally on any target of area A, the number of electrons hitting that target per second is

$$\psi^2 v A. \tag{5}$$

Having made these assumptions, the first thing we have to do is to prove that they make reasonable predictions about beams of electrons in problems where the Newtonian mechanics has proved successful. Thus a beam of electrons moving through an electric field perpendicular to its path will be bent, as illustrated in fig. 15. A beam between two grids at different potentials will be accelerated or slowed down. We have to ask, will the assumption that electron beams behave like waves give the *same* results as Newtonian mechanics? If it did not, the assumption would have to be abandoned, since Newtonian mechanics gives results in agreement with experiment for this kind of situation.

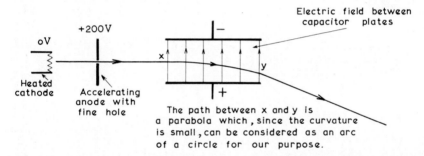

Fig. 15. Illustrating the bending of a beam of electrons between the plates of a capacitor.

The clue to our understanding of problems of this kind is that, when a beam passes through an electric field, its momentum changes from point to point and therefore the wavelength is changing too; in other words, diffraction occurs. This is best described by introducing the potential energy $V(x, y, z)$ of the electron at any point with Cartesian coordinates (x, y, z) in space. Then if W is the kinetic energy of the electron before it enters the field, where we take $V(x, y, z)$ to be zero, its kinetic energy at the point (x, y, z) is

$$W - V(x, y, z).$$

Therefore from equations (2) and (3) the wavelength λ at the point (x, y, z) is given by

$$\lambda = h/\sqrt{\{2m_{\mathrm{e}}(W - V)\}}. \tag{6}$$

33

The wavelength thus changes with position. Taking this into account, we can consider the problem illustrated in fig. 15. If a field of magnitude E is acting perpendicular to the beam, the force on each electron is eE. The path of the electron is of course a parabola but (as is usual when defining the curvature of any point on any curve), if we take a small enough element of the path, it can be regarded as part of a circle; here the radius of curvature at each point is so large in comparison with the length of the path, and the curvature changes so little, that this approximation is justified, and we can consider that the electron moves initially on a curved path with constant radius of curvature R given by

$$m_e v^2/R = eE. \tag{7}$$

We have to ask whether the wave treatment shows that the beam has just this curvature. Figure 16 shows the beam, and AA', BB' are wave fronts just a wavelength apart. AB is thus equal to a wavelength

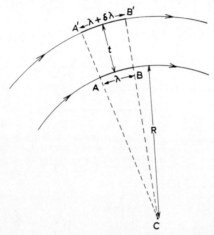

Fig. 16. Illustrating the bending of a beam of waves.

and so is A'B', but they are not equal because the wavelength varies from one point to another. If we write $AB = \lambda$, then $A'B' = \lambda + \delta\lambda$, where $\delta\lambda$ is the change in λ in going across the beam. We can determine the effective radius of curvature R from the geometry of the problem as follows. From the similar triangles CAB, CA'B' we see that

$$\frac{\lambda}{R} = \frac{\lambda + \delta\lambda}{R + t},$$

where t is the thickness of the beam, which we take to be small compared with R. Thus to the first order in small quantities

$$\frac{1}{R} = \frac{1}{t}\frac{\delta\lambda}{\lambda}.$$

34

We now remember that λ varies with position according to equation (6). Thus

$$\frac{\delta\lambda}{\lambda} = \frac{\delta V}{2(W-V)},$$

where δV is the difference between the value of V, the potential energy of an electron, at the two sides of the beam, namely eEt. It follows that

$$\frac{1}{R} = \frac{eE}{2(W-V)}. \tag{8}$$

But this is exactly the formula (7) given by Newtonian mechanics, because $W-V$ is the kinetic energy $\frac{1}{2}m_e v^2$ of an electron with total energy W. This is what we set out to prove.

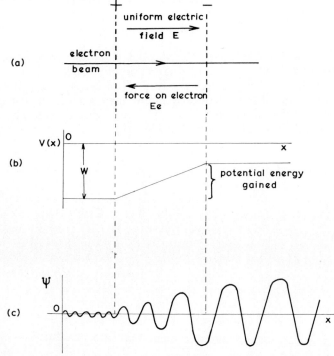

Fig. 17. Slowing down of a beam of electrons. (a) Shows the field between two grids. (b) The potential energy of an electron; W is its kinetic energy before entering the grid. (c) The wave function ψ.

The next problem that we shall look at is one in which a beam of electrons moves parallel to an electric field so that the electrons are accelerated or slowed down. An example is that of a beam of electrons, all with the initial kinetic energy W, passing through two grids between which there is a constant electric field E. Figure 17 shows (a) the grids, (b) the potential energy $V(x)$ of an electron for the case where the field decelerates the electrons, and (c) the expected form of the wave function ψ. With the field such that eE opposes the motion, the electron slows down, which means that $W - V$ in fig. 17 decreases and the wavelength increases; this is shown in the figure. The figure also shows an increase in ψ from left to right as the electrons are slowed down by the field. The physical meaning of this is as follows. The number of electrons per unit volume in the beam is, by definition, ψ^2; therefore the current, or rather the number of electrons crossing unit area per unit time, is $v\psi^2$, where v is the velocity of an electron at the point considered. But the current cannot vary with x; the phenomenon described is one in which the current is not varying with time, and there must therefore be a steady current which is the same for all values of x. Thus, since $v = \sqrt{\{2(W - V)/m\}}$ and $V(x) = eEx$, it follows that ψ ought to vary with x as

$$\psi \propto (W - eEx)^{-1/4}. \tag{9}$$

We have said that if the wave description is to describe what actually happens, namely no creation or destruction of particles when a steady stream of them passes through the field, (9) must be true. It should therefore be possible to deduce it rather than assume it, just as the bending of a beam of electrons has been deduced from the wave hypothesis. This will be carried out in § 6.2.

CHAPTER 5

standing waves† and the quantization of energy

WE have seen in the last chapter that a beam of electrons, neutrons or of other particles behaves like a beam of waves. In order to calculate the path of a beam of electrons or indeed the properties of electrons in atoms or in any other situation, it is necessary first to calculate the behaviour of this wave. Nothing is known *a priori* about this kind of wave except the experimental relationship between wavelength and momentum described in the last chapter. This chapter, therefore, begins with a description of the mathematical treatment of other kinds of wave motion, and then shows how the use of waves to predict the properties of electrons leads naturally to the quantization of energy for electrons in atoms.

We begin by introducing the 'displacement' of a wave; this is a quantity defined at every point in space, which oscillates about a mean value. In water waves the displacement is the height of the wave at any point above the average height of the surface; in a sound wave it can be the change in the pressure at any point in the wave; in a light or radio wave it is the strength of the electric or magnetic field. The simplest kind of wave, to which this discussion will refer several times, is a wave on a stretched string. In any kind of wave motion the quantity which varies with distance and time together will be denoted by the symbol ψ; in a wave on a string this quantity ψ is the displacement of the string from its mean position. The greatest value of ψ is called the amplitude A; thus, at any given point the value of ψ oscillates between the limits $\pm A$.

Fig. 18. Showing the instantaneous value of ψ at any point.

The simplest form of wave as illustrated in fig. 18 can be written

$$\psi = A \sin \{2\pi(Kx - vt)\}. \tag{1}$$

† Some books call these stationary waves. The notation here is the same as in N. F. Barber, *Water Waves* (Wykeham Publications, 1969).

37

$1/K$ is the wavelength λ, or distance from crest to crest, ν is the frequency; the displacement at any point of the wave varies with simple harmonic motion with frequency ν; that is to say, at two separated times the wave looks as in fig. 19, shown by the full and dotted lines, the crest of the wave moving forwards; after a time $1/\nu$ the crest at A has moved to B and the wave looks just as it did before. The

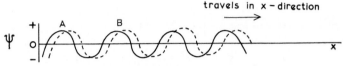

Fig. 19. Showing the form of a wave at two successive intervals of time, the dotted line being the form at a time *after* that for the full line.

'angle part' $\{2\pi(Kx-\nu t)\}$ is called the phase. The velocity, v_{wave}, with which the wave crest moves in the direction of travel of the wave, is called the 'wave velocity' or 'phase velocity' and is given by the equation

$$v_{\text{wave}} = \lambda\nu.$$

For some forms of wave motion the wave velocity is independent of frequency; this is so, for instance, for light in a vacuum, and for a stretched string, where $v_{\text{wave}} = \sqrt{(T/\varrho)}$, T being the tension and ϱ the mass per unit length. But it is not so for light in a medium where v_{wave} varies with wavelength, giving rise to *dispersion*. For light waves the ratio of the velocity c in a vacuum to the wave velocity v_{wave} in the medium is the refractive index of the medium μ, so that

$$\mu = c/v_{\text{wave}}.$$

Wave motion can occur in a medium in which the refractive index varies continuously from point to point. An example would be a cord where the mass ϱ per unit length is non-uniform. The wave velocity v_{wave} is given by

$$v_{\text{wave}} = \sqrt{(T/\varrho)},$$

so that if ϱ decreases with x the wave amplitude will appear as in fig. 20.

Fig. 20. Wave on a string in which the mass per unit length increases from left to right.

Equation (1) represents a wave going from left to right; for a wave in which the wave crests move from right to left we may write

$$\psi = A \sin \{2\pi(Kx+\nu t)\}, \tag{2}$$

38

where K, ν are the same quantities as before.   An important feature of all forms of wave motion is that if any two wave forms such as (1) and (2) are superimposed (added to each other), a possible form of the wave is obtained.   Adding (1) to (2), therefore, we obtain

$$\psi = 2A \sin 2\pi Kx \cos 2\pi \nu t. \tag{3}$$

A wave of this kind is called a 'standing or stationary wave'.

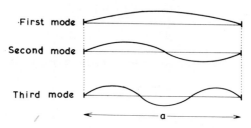

Fig. 21.   Standing waves in a string of length a, showing the first, second and third normal modes.

We see a string or piano wire secured at both ends *vibrating up and down*.   In fact, this pattern is the superposition of two sets of waves running *along* the string.   Thus a wave on a string or piano wire secured at both ends is necessarily a standing wave.   In such a case, only a series of discrete frequencies is possible.   Thus a string can vibrate in such a way that one half of a wave length is equal to the length a of the string as in fig. 21 (first mode).   Thus

$$\tfrac{1}{2}\lambda = a,$$

and the frequency is

$$\nu = v/\lambda = v/2a.$$

Other possible modes of vibration are those for which

$$\tfrac{1}{2}n\lambda = a,$$

where n is any integer, and the frequencies ν_n with which the string can vibrate are thus

$$\nu_n = nv/2a.$$

The vibrations illustrated in fig. 21 are called the 'normal modes' of the string and the frequencies ν_n the 'characteristic frequencies'.

A non-uniform string secured at the two ends will also have normal modes and characteristic frequencies, but these are not given by any simple formula.

Now let us apply these ideas about standing waves to electrons. Suppose a beam of electrons is confined in a box between reflecting planes at $x = 0$ and at $x = a$; the situation is illustrated in fig. 22. The beam of electrons going forward will be represented by a wave of the form

$$\sin\{2\pi(Kx - vt)\},$$

and the reflected wave by

$$\sin\{2\pi(Kx + vt)\}.$$

If we add these together we find as before

$$2\sin 2\pi Kx \cos 2\pi vt.$$

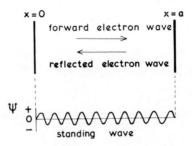

Fig. 22. Wave function of an electron shut up in a box. The forward and reflected electron waves give a standing wave.

Now the wave must have zero amplitude outside our box, because there are no electrons there; also the wave must not show a discontinuity at any point.† Therefore ψ must vanish at $x = 0$ and $x = a$. This is only possible if the wavelength has certain values, given by

$$2\pi Ka = n\pi,$$

i.e.

$$\lambda(= 1/K) = 2a/n. \tag{4}$$

In other words, the electron is described by a standing wave, with n half waves fitted in between $x = 0$ and $x = a$.

Now if we substitute the equation $\lambda = h/m_e v$ in (4), we see that the velocity of the electron is given by

$$v = nh/2m_e a \tag{5.1}$$

and its kinetic energy by

$$W = \tfrac{1}{2}m_e v^2 = n^2 h^2/8m_e a^2 \tag{5.2}$$

with $n = 1, 2, 3 \ldots$ or any other integer. So if, as wave mechanics demands, we try to describe the movement of the electron backwards and forwards by a wave, we can only do this if the kinetic energy of the electron is given by one of the quantized values (5.2). So we are led to

† This is proved in Chapter 6.

40

the conclusion that *either* the energy of an electron moving in a confined space is quantized, *or* that the wave description must be abandoned. As Chapter 1 has emphasized, the energies of electrons in atoms are quantized, and the discussion of this chapter shows why this is so. *An electron in an atom is shut up in a confined space, and the associated wave is therefore a standing wave.*

Equation (5), crude as it is as a description of an atom or a nucleus, gives us a realistic relationship between the size of an atom or a nucleus and the energies of particles in it. The diameter of an atom is a few multiples of an ångström. If in (5.1) we set $n = 1$ and a equal to 2 Å we find

$$v = 1 \cdot 5 \times 10^6 \, \mathrm{m \, s^{-1}}$$

and that this corresponds to an energy of 10^{-18} J $= 6$ eV, which is of the order of the ionization energies of an atom. Nuclei, on the other hand, have diameters of the order 10^{-14} m, and the particles are nucleons (protons and neutrons) nearly 2000 times heavier than the electron. So from (5) the velocity of the nucleon in the nucleus should be about five times higher (say 10^7 m s^{-1}). Actually alpha particles resulting from the spontaneous disintegration of nuclei do have velocities of this order. Moreover we predict that the energies involved in nuclear reactions should therefore be higher than in chemical reactions involving forces due to the atomic electrons by a factor of about $5^2 \times 2000 = 50\,000$, and thus of order 10^{-14} J $= 200\,000$ eV; this again is what is observed.

Two of the most important predictions of the Old Quantum Theory discussed in Chapter 3 were the quantization of angular momentum and of the energy of the simple harmonic oscillator. Let us see whether the standing-wave concept can explain these. We start with angular momentum and take the simple concept of a particle of mass m moving on a circle. This does not correspond closely to anything in atomic physics, since the electron in an atom is not constrained to move on a circle, but it gives an elegant and simple example of quantization. Consider the particle at P in fig. 23 (*a*), at a distance x from some fixed point O on the circle, measured along the circumference. The wave which describes it moving round the circle with velocity v is

$$\psi = \sin \{2\pi(Kx \pm vt)\}, \quad K = mv/h.$$

The plus or minus represents the fact that it can move in either direction. But if the wave function is to come back to the same value on going round the circle, so that it is single valued, an *integral* number of wavelengths must fit into the circumference; otherwise there will be a misfit as in fig. 23 (*b*). It follows that

$$l\lambda = \text{the length of the circumference}$$
$$= 2\pi r,$$

where r is the radius of the circle, and l is an integer. Writing for the particle $\lambda = h/mv$, this gives

$$mvr = lh/2\pi, \tag{6}$$

which is Bohr's condition for the quantization of angular momentum.

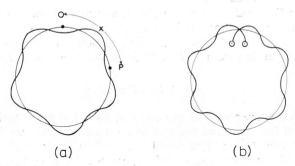

(a)   (b)

Fig. 23. Wave function of a particle on a circle. (a) A continuous single-valued solution of the equation which does not describe a physical state. (b) A solution which does not describe a physical state.

There is however one difference between this condition and that postulated by Bohr; there is no reason why l should not have the value zero in equation (6). The angular momentum can therefore be zero. The state of lowest angular momentum, therefore, has zero energy. The excited states of the system, with $l = 1, 2, 3 \ldots$ are what is known as 'degenerate'. That means that, for every allowed value of the energy,

$$\tfrac{1}{2}mv^2 = l^2h^2/8\pi^2mr^2, \tag{7}$$

there are *two* possible wave functions, corresponding to the direction of v in equation (6). These energy levels are shown in fig. 24.

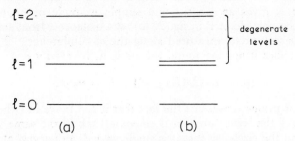

(a)   (b)

Fig. 24. Energies of stationary states of an electron moving on a circle. (a) Without a magnetic field. (b) With a magnetic field.

In the case of electrons or other charged particles, if there is a magnetic field perpendicular to the plane of the circle, the energies of all the degenerate states, except that for which $l = 0$, are split into pairs

42

as shown in fig. 24 (*b*). The reason is that an electron moving in a circular orbit produces a magnetic moment μ, and the direction of the moment depends on whether the electron is moving clockwise or anticlockwise. So the energy of the moment in a magnetic field B is $\pm B\mu$; $\pm B\mu$ has to be added to the original energy.

We shall now calculate μ. A charge e crosses any point on the circumference of the circle $v/2\pi r$ times in each second, so we can say that a current $j = ev/2\pi r$ is flowing in 'the wire'. The question is, what magnetic moment does a current of this magnitude produce? The answer is

current × *area of circle.*

So the moment is

$$\pi r^2 \times ev/2\pi r,$$

which reduces to

$$\mu = \tfrac{1}{2}evr.$$

If we substitute from (6), which gives the quantization condition for the angular momentum, we obtain, setting $m = m_e$

$$\mu = l\mu_0,$$

where

$$\mu_0 = eh/4\pi m_e.$$

μ_0 is known as the Bohr magneton, and is a 'quantum' of magnetic moment.† Its magnitude is $0.974 \times 10^{-23} \, \text{JT}^{-1}$. The simple theory thus predicts that the magnetic moment must be an integral multiple of this quantum.

According to the direction of the current, then, the energy due to the magnetic moment in the field B is

$$\pm l\mu_0 B.$$

So, unless $l = 0$, the states of an electron are split as illustrated. There is a state with energy

$$l^2 h^2/8\pi^2 m_e r^2 + l\mu_0 B$$

and one with energy

$$l^2 h^2/8\pi^2 m_e r^2 - l\mu_0 B.$$

A state which can be split in two or more states in this way is called 'degenerate'. We see that the state with $l = 0$ is not degenerate.

A magnetic field B, then, splits the energy levels of a degenerate state into a number of states separated by an energy

$$\Delta W = \mu_0 B.$$

† In the electrostatic units in which the concept was originally formulated, we should write $\mu_0 = eh/4\pi m_e c$, where c is the velocity of light.

43

Therefore spectral lines should be split by multiples of a frequency $\Delta \nu$ where

$$\Delta \nu = B\mu_0/h = Be/2m_e. \tag{8}$$

Such a splitting is observed, and is known as the Zeeman effect, though in many cases the behaviour is more complicated than that described here. This splitting was discovered before the introduction of quantum mechanics, and the formula (8) above (which does not contain h) was obtained using a classical model. The explanation of a feature of line spectra in terms of a formula involving the quantity e/m_e, previously determined for cathode rays, was one of the earliest pieces of evidence which showed that electrons are involved in the emission of radiation by atoms.

We turn next to the problem of the simple harmonic oscillator. The problem, outlined in Chapter 2, is that of a particle of mass m moving on a straight line, and subject to a restoring force qx when displaced a distance x from its mean position. Such a particle will vibrate with frequency ν given by

$$\nu = \frac{1}{2\pi} \sqrt{\left(\frac{q}{m}\right)}. \tag{9}$$

At displacement x, its potential energy is $\frac{1}{2}qx^2$, and if the total energy is W, the kinetic energy is

$$W - \tfrac{1}{2}qx^2.$$

We have here, then, a problem in which the velocity, and therefore the wavelength of the wave, varies with position. Suppose however we overlook this for the time being, and take the wavelength to be that at $x = 0$ when all the energy is kinetic, namely

$$\lambda = h/\sqrt{(2mW)} \tag{10}$$

(see Chapter 4, equation (3)). The particle is moving backwards and forwards over a distance $2x_0$ where

$$\tfrac{1}{2}qx_0{}^2 = W, \tag{11}$$

because when $x = x_0$ all the energy is potential energy and the particle is at rest. For a standing wave we have to fit n half waves into this distance, so that

$$\tfrac{1}{2}n\lambda = 2x_0. \tag{12}$$

Using equations (9), (10) and (11) to eliminate x_0, λ and expressing q/m in terms of ν through (9), this gives

$$W = (\pi/4)nh\nu,$$

where $n = 1, 2, 3. \ldots$ This is quite near the correct answer, which is $W = (n + \frac{1}{2})h\nu$ $(n = 0, 1, 2 \ldots)$. With the approximations we have

44

made we would not expect anything better. We shall return to this problem in the next chapter.

5.1 *The complex form of the wave function*

Up to this point we have described the wave function ψ of a progressive wave by the equation

$$\psi = A \sin \{2\pi(Kx - vt)\} \tag{13}$$

and a standing wave by

$$\psi = A \sin 2\pi Kx \cos 2\pi vt. \tag{14}$$

In quantum mechanics we prefer to use another way of writing these equations, and take instead for a progressive wave

$$\psi = A \exp \{2\pi i(Kx - vt)\} \tag{15}$$

and for a standing wave

$$\psi = A \sin 2\pi Kx \exp(-2\pi i vt), \tag{16}$$

where $i = \sqrt{(-1)}$. The reasons for this will now be explained.

First let us look at the behaviour of complex quantities in general. The quantity $\exp(i\alpha)$, whatever α may be, is by de Moivre's theorem given by

$$\exp(i\alpha) = \cos \alpha + i \sin \alpha.$$

Any complex quantity ψ can be written in the form

$$\psi = A \exp(i\alpha),$$

where A and α are real. The quantity A is called the 'modulus' of ψ and is written $|\psi|$.

If we write the wave function ψ in the form (15) or (16), and take $|\psi|^2 dx$ as the probability that a particle will be found at a point between x and $x + dx$, then we see that this probability does not vary with time, which is what we should expect. If we used the form (14), then the quantity $\psi^2 dx$ over the length of any confined space in which the electron is supposed to be would oscillate between zero and some maximum value, suggesting that the particle keeps disappearing! This is of course a nonsensical prediction, and is one reason why quantum mechanics has been developed using the complex form for ψ.

It is sometimes thought surprising that a wave function ψ, which is related to observable physical properties, should have a complex form. In a sense the use of the complex form is just a convenience. If one writes

$$\psi = f + ig,$$

where f and g are *real* functions of position and time, then the important point is that both f *and* g are needed to specify the behaviour of the

45

wave.† In this respect, de Broglie waves differ in no respect from any other wave system. Thus for a wave on a string, if y is the displacement and $\dot{y}$ the velocity of any point of the string, one needs to know both y and $\dot{y}$ if the future motion of the string is to be determined. Moreover, the energy of any unit length of the string is the sum of two squared terms, the potential energy $\frac{1}{2}Ty^2$ and the kinetic energy $\frac{1}{2}\varrho\dot{y}^2$, T being the tension and ϱ the mass per unit length. For the electron wave,

$$|\psi|^2 = f^2 + g^2$$

gives the chance per unit length that an electron will be found at any point, and there is a certain analogy between the probability density for the electron wave and the energy density for a wave on a string.

5.2 *Normalization of the wave function*

For the standing wave representing a particle confined between planes at $x = 0$ and $x = a$, instead of the notation of the last chapter, the wave function should be

$$\psi = A \sin(\pi n x/a) \exp(-2\pi i \nu_n t) \tag{17}$$

with $W_n = n^2 h^2/8ma^2$. The constant A must be chosen so that

$$\int_0^a |\psi|^2\, dx$$

is equal to the number of particles confined between $x = 0$ and $x = a$. If there is just one particle this gives

$$A^2 = 2/a.$$

If A is chosen in this way, the wave function (17) is said to be normalized.

---

† The same is true for the complex conjugate $\psi^* = f - ig$. The product $|\psi|^2 = \psi\psi^* = (f+ig)(f-ig) = f^2 + g^2$ gives the intensity of the wave.

the Schrödinger equation

THE last chapters have shown how the observed relationship between the velocity v of an electron and the wavelength $h/m_e v$ of the wave describing it makes it possible to understand why the energy of an electron confined in a limited space is quantized. In order to obtain quantitative results in all but the simplest examples, however, one must make use of the differential equation satisfied by the wave function ψ; this is the Schrödinger equation, introduced into physics by the German physicist Erwin Schrödinger in 1926. This we shall now derive.

For progressive waves we have written

$$\psi = A \exp \{2\pi i(Kx - vt)\},$$

where K, the wave number, is equal to $1/\lambda$. For standing waves we have written

$$\psi = A \sin 2\pi Kx \exp (-2\pi i vt).$$

In either case ψ varies with x, for a fixed value of t, in such a way as to satisfy the simple harmonic equation

$$\frac{d^2\psi}{dx^2} + 4\pi^2 K^2 \psi = 0, \tag{1}$$

as may be verified by direct differentiation, the time t being treated as a constant. If a field is present, so that the potential energy $V(x)$ of an electron varies with x, K will be a function of x too. From the experiments on electron diffraction we know that $K = m_e v/h$. Thus

$$K^2 = m_e^2 v^2/h^2$$

$$= 2m_e(W - V)/h^2, \tag{2}$$

where W is the total energy of each electron, so that $W - V$ is the kinetic energy. Substituting (2) into equation (1), we see that the wave function which describes the behaviour of an electron must satisfy the differential equation

$$\frac{d^2\psi}{dx^2} + \frac{8\pi^2 m_e}{h^2} \{W - V(x)\}\psi = 0. \tag{3}$$

This is Schrödinger's equation.

To illustrate the properties of a differential equation of this kind we consider a specific case, that of an electron moving in an electric field E, so that the potential energy at distance x from the zero of potential energy is eEx. The quantities $V(x)$ and W are shown in fig. 17 of

Chapter 4. The argument, however, will apply to any equation of the form

$$\frac{d^2\psi}{dx^2} + F(x)\psi = 0, \tag{4}$$

where $F(x)$ is some known function.

We first show that, if ψ and $d\psi/dx$ are given for some value of x, equation (4) enables ψ to be drawn as a function of x. In fig. 25 the value of ψ for a certain value of x, shown as ON, is represented by PN, and the tangent to the plot of ψ as function of x by the line AP. Then $d^2\psi/dx^2$ is equal to the rate of change of $\tan\theta$, where θ is the angle between the tangent and Ox. So if we go along the curve to a new point P', the quantity

$$\frac{d^2\psi}{dx^2} \text{ multiplied by the length NN}'$$

is equal to the change in the angle θ. We can thus draw a new tangent at P'; and the process can be repeated, and in this way the curve can be traced out.

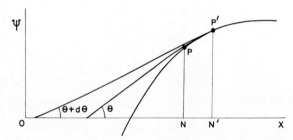

Fig. 25. A function ψ plotted against x to show the numerical solution of equation (4). Note that $d\theta$ is negative, as it will be if $F(x)$ is positive.

If, by starting with two different initial conditions two curves are drawn,

$$\psi = \psi_1(x)$$

and

$$\psi = \psi_2(x),$$

then *all* solutions of the equation are of the form

$$\psi = A\psi_1(x) + B\psi_2(x).$$

This follows because we can choose A and B so that both ψ and $d\psi/dx$ for a given value of x have any values that we like to impose.

A third point about equation (4) is the following: if $F(x)$ is positive, $d^2\psi/dx^2$ always has opposite sign to ψ. This means that the plot of ψ

48

against x is always concave to the x axis, as in fig. 25, and therefore ψ is an oscillating function, like a sine curve but with a changing wavelength. But if $F(x)$ is negative, ψ is convex to the x axis, and quite a different situation arises. What happens can be most easily seen if $F(x)$ is a negative constant; we may then set

$$F(x) = -\gamma^2$$

and (4) becomes

$$\frac{d^2\psi}{dx^2} - \gamma^2\psi = 0.$$

The two independent solutions are

$$\psi = \exp(\gamma x), \quad \psi = \exp(-\gamma x).$$

These behave quite differently from the solutions for the case where $F(x)$ is positive, as shown in fig. 26; the first solution, curve (a), tends to infinity as x increases, the other (b) tends to zero. Both are convex to the x axis. There will be two independent solutions of this kind, whether $F(x)$ is constant or not.

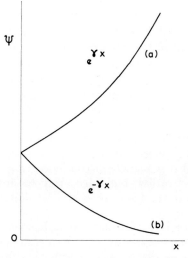

Fig. 26. The two independent solutions of equation (4) when $F(x)$ is negative. (a) The unbounded solution. (b) The bounded solution.

We now apply this analysis to the situation illustrated in fig. 17, in which a beam of electrons is moving in an electric field which slows them down. In Newtonian mechanics, the electrons will gradually be slowed down until they come to the point where the potential energy is equal to the total energy, so that x is given by

$$eEx = W;$$

49

they are then reflected. A standing wave with gradually increasing wavelength up to this point is to be expected. But beyond it, the wave function does not suddenly vanish; any solution of equation (3) must have one of the forms illustrated in fig. 26. But a solution which *increases* with increasing x is physically meaningless; the solution which represents the reflection of a wave is that which has the form (b) in the figure, dying away exponentially as x increases into the region where $W - V$ is negative. In choosing this solution of the equation, the one which dies away, we are choosing a unique solution for all x, including the oscillating part. In fig. 17 the condition that the wave is reflected determines the position of the zeros, just as much as it does when a wave is reflected from a rigid barrier.

Quantum mechanics, then, leads to a new and important result. When a beam of particles is reflected by a field which slows them down till the velocity is zero, there is a small probability (represented by $|\psi|^2$) that the particle goes a little further than the classical particle can go. It can 'tunnel' into the region in which its Newtonian kinetic energy is negative. This prediction of quantum mechanics is known as the tunnel effect, and its consequences are explored later in this chapter.

6.1 *Some problems involving potential steps*

In order to illustrate the properties of the wave function that have just been described, we shall look at some problems involving beams of particles incident on 'potential steps'. By a potential step we mean a plane, perpendicular to the motion of the particle, at which there is a discontinuous change in its potential energy, as illustrated in fig. 27 (*a*). An approximation to this situation occurs if there is a very strong field extending over a very short distance, such as exists at the surface of a metal. A change that is actually discontinuous does not correspond to anything in physical reality, but since problems involving such steps allow of exact solutions, they are useful to illustrate the principles involved.

Our first problem is to calculate what happens when a beam of particles, each with kinetic energy W, is incident on a plane where there is a potential step such that U is less than W. Since there is a *sharp* change in the potential, and thus a sharp change in the wavelength of an electron wave, the wave should behave like a light wave incident on a slab of glass; we should expect it to be partly reflected and partly transmitted. The amounts transmitted and reflected can be calculated as follows. Taking the potential energy $V(x)$ to be

$$V = U, \quad x > 0$$
$$V = 0, \quad x < 0,$$

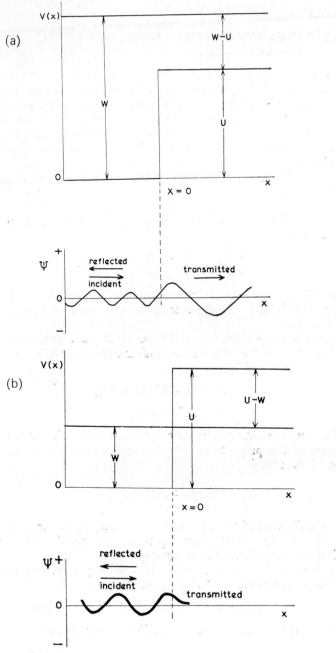

Fig. 27. A potential step, showing the wave function for an incident and reflected wave.   (a) $W > U$.   (b) $W < U$.

51

we suppose the wave function to be

$$\psi = [\exp(2\pi iKx) + A \exp(-2\pi iKx)] \exp(-2\pi i\nu t), \quad x < 0$$
$$= \exp(2\pi iK'x - 2\pi i\nu t), \qquad\qquad x > 0,$$

where $K^2 = 2mW/h^2$, so that $K = 1/\lambda = mv/h$, $K'^2 = 2m(W-U)/h^2$.
The wave function represents one particle per unit volume in the incident wave, $|A|^2$ particles per unit volume in the reflected wave and $|B|^2$ in the transmitted wave. Thus v particles cross unit area per unit time in the incident wave, $v|A|^2$ in the reflected and $v'|B|^2$ in the transmitted wave. Since $v'/v = K'/K$, the proportion of the beam reflected is $|A|^2$ and the proportion transmitted is $(K'/K)|B|^2$.

Our problem is to calculate A and B. For this we need to know the boundary conditions satisfied at $x = 0$. These are, that ψ and $d\psi/dx$ should be continuous; there is no kink in ψ (fig. 27). This may be proved as follows. From the Schrödinger equation, integrating both sides we find

$$\frac{d\psi}{dx} = -\frac{8\pi^2 m}{h^2} \int_0^x (W-V)\psi \, dx.$$

Even if V is discontinuous, its integral must be continuous as one can see from fig. 27 (a). So $d\psi/dx$ is continuous and therefore ψ must be continuous too.

Putting in these boundary conditions, since ψ is continuous at $x = 0$ it follows that $1 + A = B$ and since $d\psi/dx$ is continuous $K(1-A) = K'B$. Solving for A and B,

$$A = (K-K')/(K+K'),$$
$$B = 2K/(K+K').$$

The most interesting thing to be deduced from this solution is that the sum of the proportions reflected and transmitted comes out to unity. This is so if

$$|A|^2 + \frac{K'}{K}|B|^2 = 1.$$

This is easily verified to be the case. If it were not so, there would be something wrong with the wave equation, as it would predict creation or disappearance of particles at the step.

We turn now to the case when $U > W$ (fig. 27 b); we expect all particles to be reflected. As before we set, for $x < 0$,

$$\psi = [\exp(2\pi iKx) + A \exp(-2\pi iKx)] \exp(-2\pi i\nu t).$$

For $x > 0$ the general solution of the Schrödinger equation is

$$[B \exp(-2\pi\gamma x) + C \exp(2\pi\gamma x)] \exp(-2\pi i\nu t), \quad \gamma^2 = \frac{2m(U-W)}{h^2}.$$

For the reasons given in the last section, to describe total reflection we must take the solution that *decreases* with increasing x, and thus set $C = 0$. Thus we have

$$\psi = B \exp\left(-2\pi\gamma x - 2\pi i\nu t\right), \quad x > 0.$$

Putting in the boundary conditions as before we find

$$1 + A = B,$$

$$iK(1 - A) = -\gamma B.$$

Eliminating B we have

$$A = \frac{iK + \gamma}{iK - \gamma}. \tag{5}$$

The important thing here to note is that $|A| = 1$, so the reflected wave has the same amplitude as the incident one. This will be obvious to the reader familiar with complex numbers; to others it can be shown as follows. One can always write any complex number $\gamma + iK$ as

$$\gamma + iK = r \exp(i\alpha),$$

where r and α are real and $\alpha = \tan^{-1}(K/\gamma)$; the equation gives $r \cos \alpha = \gamma$, $r \sin \alpha = K$, so that $r = \sqrt{(\gamma^2 + K^2)}$ and $\alpha = \text{arc tan}\,(K/\gamma)$. So A in (5) is of the form $\exp(2i\alpha)$, and the modulus of this is unity. The incident and reflected wave are thus denoted by

$$\psi = [\exp(2\pi iKx) + \exp(-2\pi iKx + 2i\alpha)] \exp(-2\pi i\nu t)$$

$$= \exp(i\alpha) \cdot 2\cos(2\pi Kx - \alpha) \exp(-2\pi i\nu t).$$

This example shows us that the wave penetrates into the region where $x > 0$, where the kinetic energy is negative and the classical particle cannot go. This is an example of the 'tunnel effect'. ψ will vanish at $x = 0$ only if U is infinite, as a simple calculation shows. The first of these expressions shows that ψ represents a standing wave, waves of equal amplitude (equal to unity) moving in the two directions. The factor $\exp(i\alpha)$ outside the second expression is arbitrary; by this is meant that one could multiply both expressions by $\exp(-i\alpha)$ without affecting the physical interpretation of the wave function in terms of density of particles.

The penetration of a particle beyond the plane $x = 0$ leads us to investigate a 'potential barrier' illustrated in fig. 28. If particles are incident on such a barrier, most will be reflected but some will penetrate through it, as we shall see.

We set

$$V(x) = 0, \quad x < 0$$

$$= U, \quad 0 < x < a$$

$$= 0, \quad x > a.$$

53

Then (omitting the factor $\exp(-2\pi i \nu t)$ from all equations)

$$\psi = \exp(2\pi i K x) + A \exp(-2\pi i K x) \quad x < 0$$
$$= B \exp(2\pi \gamma x) + C \exp(-2\pi \gamma x) \quad 0 < x < a$$
$$= D \exp[2\pi i K(x-a)]. \quad x > a.$$

The boundary conditions are at $x = 0$

$$1 + A = B + C,$$
$$iK(1-A) = \gamma(B-C)$$

and at $x = a$

$$B \exp(2\pi\gamma a) + C \exp(-2\pi\gamma a) = D,$$
$$\gamma[B \exp(2\pi\gamma a) - C \exp(-2\pi\gamma a)] = iKD.$$

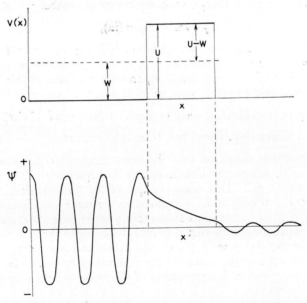

Fig. 28. A potential barrier, showing the wave function of an electron tunnelling through it.

The elimination of A, B and C is straightforward but laborious. One finds

$$D = \frac{-4i\gamma K}{(\gamma^2 + K^2)[\exp(2\pi\gamma a - 2i\theta) - \exp(-2\pi\gamma a + 2i\theta)]}.$$

54

where $\tan \theta = K/\gamma$. If one has a thick barrier, so that $\exp(2\gamma a)$ is large, the second term in the square bracket can be neglected. We then have

$$|D|^2 = \frac{16\gamma^2 K^2}{(\gamma^2 + K^2)^2} \exp(-4\pi a\gamma).$$

This gives the *proportion* of all particles incident on the beam which can penetrate it. The important term is $\exp(-4\pi a\gamma)$; the thicker the barrier and the greater the barrier height $U - W$, the fewer particles will get through.

Example

Calculate the factor $\exp(-4\pi a\gamma)$ for electrons when the barrier is 10 Å thick, $U = 11$ eV and the kinetic energy of each electron is 1 eV. The quantities K, K' are the wave-numbers (the reciprocal of the wavelength) for the wave to the left and to the right of the potential step; in terms of the velocity v, $K = m_e v/h$ and in terms of the energy

$$K^2 = 2m_e W/h^2, \quad K'^2 = 2m_e(W - U)/h^2.$$

6.2 *The method of Wentzel, Kramers and Brillouin*

This is a useful method of obtaining approximate solutions of the Schrödinger equation or of the equation (4, p. 48), in cases where the function $F(x)$ does not vary much in one wavelength. The solution is expected to be of the form illustrated in fig. 20, namely a function with slowly varying wavelength and slowly varying amplitude. We write it therefore

$$\psi = A(x) \exp\{iB(x)\}, \tag{6}$$

where A and the *rate of change* of B are small and substitute (6) in the equation (4). We have, using ψ' to denote $d\psi/dx$, etc.,

$$\psi' = (A' + iAB') \exp(iB)$$
$$= \{A'' - AB'^2 + i(2A'B' + AB'')\} \exp(iB). \tag{7}$$

What we do now is to neglect A''. This means that, though A is changing with x, the rate of change and particularly the change in the rate of change is small. If, neglecting A'', we substitute (7) into the equation (4), we obtain

$$\{-AB'^2 + AF(x)\} + i(2A'B' + AB'') = 0. \tag{8}$$

Now a solution to an equation such as (8) can be obtained by putting the real and imaginary parts equal to zero. Let us take the first. This gives

$$B = \int_0^x \{F(x)\}^{1/2} \, dx.$$

If F is a constant, this gives $F^{1/2}x$, which we know already to be the right solution. If F is varying slowly, then over a number of wavelengths it behaves like $F^{1/2}x + $ const., so the solution does represent a wave of varying wavelength, which is just what we want.

Of greater interest is the amplitude A. This is obtained by equating to zero the imaginary part of the equation (8). Thus

$$AB'' + 2A'B' = 0.$$

This can be written

$$\frac{B''}{B'} + \frac{2A'}{A} = 0$$

and can be integrated to give

$$\log B' + 2 \log A = \text{const.}$$

or

$$\log (A^2 B') = \text{const.}$$

This gives for A

$$A = \text{const.} \, (B')^{-1/2}.$$

Since $B' = F^{1/2}$, we find

$$A = \text{const.} \, F^{-1/4}.$$

In terms of wave mechanics, this has the following interpretation. In Schrödinger's equation (3), the quantity F is $8\pi^2 m(W - V)/h^2$, so

$$A^2 = \text{const.}/\sqrt{(W - V)}.$$

$W - V$ is the kinetic energy of the particle, so that

$$A^2 = \text{const.}/v,$$

where v is the velocity with which the particle moves.

We have already pointed out on p. 36 that this must be so if quantum mechanics is to make sense in its description of a beam of particles. As they slow down in an electric field, they come closer together, and $A^2 v$ must be constant along the beam. The analysis we have just given, then, shows that wave mechanics at any rate gives a result which makes sense and does not predict the creation of particles.

6.3 The application of the method of Wentzel, Kramers and Brillouin to alpha-decay

As we have stated already, the wave function ψ does not vanish in the region where $W - V$ is negative, which is where the particle cannot go according to Newtonian mechanics. The analysis given above can be adapted very simply to determine the behaviour of ψ in this region. Our Schrödinger equation is now of the form

$$\frac{d^2\psi}{dx^2} - G(x)\psi = 0,$$

where G is positive. Instead of (6) we write

$$\psi = A \exp(-B), \tag{9}$$

and obtain by exactly the same kind of analysis

$$B'^2 = G(x),$$

so that

$$B = \int_0^x \{G(x)\}^{1/2} \, dx. \tag{10}$$

The most famous example of quantum mechanical tunnelling to be treated in this way is the explanation, given first by the Russian physicist George Gamow in 1928, of the decay of those of the radioactive nuclei which emit alpha particles. The alpha particle, which is a helium nucleus with charge $2e$ and mass M about 7000 times that of the electron, is emitted with energies of between one and two million electron volts—say 10^5 times the ionization energy of hydrogen. The de Broglie wavelength for a particle of mass M and kinetic energy W is

$$h/\sqrt{(2MW)},$$

and we should therefore expect it to be about $(7000 \times 10^5)^{1/2}$ less than for an electron in a hydrogen atom ($\sim 10^{-10}$ m) and thus about 0.4×10^{-14} m. This is about the radius of an atomic nucleus, and it is thus entirely possible that the alpha particle moves backwards and forwards in an atomic nucleus in very much the same way that an electron does in a hydrogen atom.

We shall not discuss the nature of the forces which bind together the neutrons and protons which make up a nucleus. They are not fully understood now, and almost nothing was known about them at the time of Gamow's work. Gamow assumed that there must be *some* attractive force, setting in at a distance equal to the radius of the nucleus so as to hold the alpha particle in.

But once the alpha particle is clear of the nucleus, its potential energy is simply the Coulomb term $2(Z-2)e^2/4\pi\varepsilon_0 r$ due to the repulsion of the nucleus, which is taken to have charge Ze before the alpha particle escapes. Thus the whole potential energy curve is as shown in fig. 29, BC being the Coulomb part.

Now we have to ask, under what conditions will an alpha particle get out? If the energy is negative, as represented by the line XY in fig. 29, it never will; it would have to gain energy to escape, and this can only happen if it is hit hard by another particle. But if the energy is positive, as shown by the line X'Y', then according to Newtonian mechanics it cannot get out, but according to quantum mechanics it does so by 'tunnelling' through the barrier between P and Q. Moreover, using (10) with $G = 8\pi^2 M\{V(r) - W\}/h^2$, shows that between P and Q the wave function decreases by the factor

$$\psi_Q/\psi_P = \exp\left[-\int_P^Q \frac{8\pi^2 M}{h^2}\{V(r) - W\}^{1/2}\,dx\right] \qquad (11)$$

and $|\psi|^2$, the amount by which the chance of finding a particle drops, by the square of this. The particle has a velocity of about $10^7\,\mathrm{m\,s^{-1}}$, it moves backwards and forward in a nucleus of diameter about 10^{-14} m, so it tries to get out 10^{21} times per second.

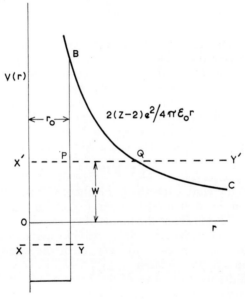

Fig. 29. Potential energy of an alpha particle in neighbourhood of nucleus. The lines XY, X'Y' represent the energies for a stable and unstable nucleus.

The chance per second that it gets out, which we call λ, is given by

$$\lambda = 10^{21}|\psi_Q|\psi_P|^2\,\mathrm{s^{-1}}.$$

This quantity λ is the decay constant. If there are N nuclei, the number decaying in time dt is

$$\lambda N\,dt$$

so the rate of change of N is given by the equation

$$\frac{dN}{dt} = -\lambda N$$

58

and on integrating we find $N = N_0 \exp(-\lambda t)$, where N_0 is the initial value of N.

The integral in (11) can be evaluated if we take for $V(r)$ the simplified form (writing $Z - 2 = Z'$)

$$V(r) = 2Z'e^2/4\pi\varepsilon_0 r \quad r > r_0$$

$$= \text{const.} \quad r < r_0,$$

where r_0 is a measure of the nuclear radius. $|\psi_Q|\psi_P|^2$ is thus

$$\left|\frac{\psi_Q}{\psi_P}\right|^2 = \exp\left[-2\int_{r_0}^{2Z'e^2/4\pi\varepsilon_0 W}\left\{\frac{8\pi^2 M}{h^2}(V(r) - W)\right\}^{1/2} dr\right].$$

The integral can be evaluated by setting

$$\cos^2 u = r/r_1,$$

where $r_1 = 2Z'e^2/W$. It comes out to give

$$\ln\left|\frac{\psi_Q}{\psi_P}\right|^2 = \exp\left[-\left\{\frac{8\pi^2 M(2Z'e^2)r_1}{h^2 4\pi\varepsilon_0}\right\}^{1/2}(2u_0 - \sin 2u_0)\right],$$

where $\cos^2 u_0 = r_1/r_0$. Since r_1/r_0 is small we may write

$$u_0 = \tfrac{1}{2}\pi - \sqrt{(r_1/r_0)}.$$

The decay constant is thus, in s^{-1},

$$\sim 10^{21} \exp\left[-\frac{2\pi^2}{h}\left\{\frac{2\pi M r_1(2Z'e^2)}{\varepsilon_0 h^2}\right\}\left(1 - \frac{4}{\pi}\sqrt{\frac{r_1}{r_0}}\right)\right].$$

The quantity M is here the mass of the alpha particle. Its large value makes the exponential term extremely small, turning the factor 10^{21} s^{-1} into 10^{-12} s^{-1} or so which is the decay constant of uranium. Moreover, we see that r_0 has quite a big effect on λ, and a comparison of this formula with observed values of λ was one of the first ways in which estimates of the nuclear radii were obtained.

6.4 The Schrödinger equation in three dimensions

The problems that have been discussed up to this point relate to the movement of a particle in a single direction, such as the x axis, so that the wave function and the potential energy depend on x only. In general the wave function ψ describing the state of the particle is a function of x, y and z; the interpretation is that

$$|\psi(x, y, z)|^2 \, dx \, dy \, dz$$

gives the probability that a particle will be found in the volume element $dx \, dy \, dz$. A plane wave, moving in the direction specified by direction cosines l_1, l_2, l_3 takes the form

$$\psi = A \exp[2\pi i\{K(l_1 x + l_2 y + l_3 z) - \nu t\}]. \tag{12}$$

The generalization of Schrödinger's equation to three dimensions is

$$\nabla^2\psi + \frac{8\pi^2 m}{h^2}(W - V)\psi = 0.$$

Here ∇^2 denotes the operator

$$\nabla^2 \equiv \frac{\partial^2}{\partial x^2} + \frac{\partial^2}{\partial y^2} + \frac{\partial^2}{\partial z^2}.$$

It can easily be verified that (12) satisfies (1), making use of the relationship between the direction cosines which gives

$$l_1^2 + l_2^2 + l_3^2 = 1.$$

We shall use this equation in the next chapter.

CHAPTER 7

calculation of quantized energies

ONE of the main achievements of quantum mechanics is its ability to explain why within an atom the energy of an electron (or of a system of electrons interacting with each other) is limited to a series of discrete values. As applied to a single particle, as for instance an electron in the hydrogen atom, what quantum mechanics shows is that the possible values of the energy of an electron in a confined space are limited in this way. In Chapter 5 we showed why this is so for a particle moving along a straight line and confined between fixed points. This derivation did not use Schrödinger's equation; in this chapter we shall start our discussion of quantization by applying his equation to this simple one-dimensional example, and then turn to some more realistic problems.

We must first state, however, that the fact established by the experiments described in Chapter 1 is that the total energy of *all* the electrons in an atom or molecule is quantized. Since these electrons repel one another, one cannot say in any exact sense that each electron has a definite energy; part of the energy of the system is the potential energy due to this repulsion, and does not belong to one electron rather than to another. To calculate the energy of an atom or molecule containing more than one electron one needs a 'many-electron' theory, which will be described briefly in Chapter 9. The analysis of this chapter is applicable only to systems containing a single electron, such as the hydrogen atom or the hydrogen molecular ion H_2^+.

7.1 An electron moving on a straight line between two fixed points

The Schrödinger equation is

$$\frac{d^2\psi}{dx^2} + \frac{8\pi^2 m_e W}{h^2}\,\psi = 0 \tag{1}$$

and if the electron can move between the points $x = 0$ and $x = a$, then the arguments of the last chapter show that ψ must vanish when $x = 0$ and when $x = a$. The solution that vanishes at $x = 0$ is

$$\psi = A \sin 2\pi K x, \tag{2}$$

where K is the wave number ($K = 1/\lambda$) and since, by de Broglie's relationship $\lambda = h/\sqrt{(2m_e W)}$,

$$K^2 = 2m_e W/h^2.$$

61

This solution will vanish also when $x = a$ if and only if

$$K = \tfrac{1}{2}n/a,$$

where n is any integer ($n = 1, 2, 3 \ldots$). If K has one of these values, the energy W of the electron is given by

$$W = n^2 h^2/8 m_e a^2. \qquad (3)$$

The solutions are illustrated in figs. 21, 22.

We are thus led to the conclusion that the behaviour of a particle in a confined space can be described by a wave only if its energy has one of these quantized values; if it were found experimentally that the energy was not quantized in this way, quantum mechanics would have to be abandoned. The quantization of energy values, therefore, follows naturally from the assumption that the behaviour of the particle must be described by a wave.

If we multiply (2) by the factor containing the time, so that

$$\psi = A \sin (\pi n x/a) \exp (- 2\pi \mathrm{i} \nu_n t),$$

the nature of ψ as a standing wave becomes apparent; it can be written

$$\psi = \frac{A}{2\mathrm{i}} \left[\exp \left\{ \mathrm{i} \left(\frac{\pi n x}{a} - 2\pi \nu_n t \right) \right\} - \exp \left\{ \mathrm{i} \left(- \frac{\pi n x}{a} - 2\pi \nu_n t \right) \right\} \right].$$

The first term describes a particle moving from left to right, the second term one moving from right to left, so ψ describes a particle moving backwards and forwards along a line and being reflected at the two ends.

The probability that the particle would be found between the points x, $x+dx$ is

$$|\psi|^2 \, dx = A^2 \sin^2 (\pi n x/a) \, dx.$$

Since the particle must lie somewhere between the points $x = 0$ and $x = a$, this probability integrated between these two limits must be unity. Thus

$$A^2 \int_0^a \sin^2 (\pi n x/a) \, dx = 1.$$

A function satisfying this property is said to be normalized. Evaluation of the integral gives $A = (2/a)^{1/2}$, a result obtained in Chapter 5.

7.2 An electron moving between two grids

An infinite potential step at which ψ vanishes is of course an idealization. A more realistic example would be to consider an electron moving in the equipotential space of a positively charged cylinder between two negative grids, the two fields between grids and cylinder being in opposite directions so as to keep an electron in the space between them. The potential energy of the electrons is illustrated in fig. 30. The behaviour of the electron will be described by a wave function representing an electron moving backwards and forwards between the two grids, reflected at both of them at the points AB but 'tunnelling' beyond them as illustrated in fig. 30.

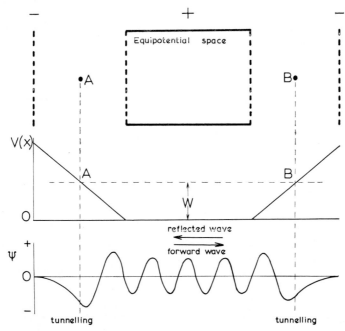

Fig. 30. Potential energy $V(x)$ and wave function ψ for an electron moving between two grids. W is the total energy, and the electron is reflected at the points A, B.

Here too solutions only exist for quantized values of the energy, though no simple formula can be obtained; for other values, a solution cannot be found which decays exponentially at *both* boundaries.

A wave function with no zeros corresponds to the ground state. The first excited state has one zero, and so on. If the states are numbered with quantum numbers n ($n = 1, 2, 3 \ldots$), then $n-1$ denotes the number of zeros in the wave function. This is true generally for problems in one dimension.

7.3 *The harmonic oscillator*

This analysis leads to a problem of great physical importance, the simple harmonic oscillator already discussed in Chapters 2 and 5; the problem is to find the allowed energy levels for any kind of particle P of mass m which is attracted to a fixed point O by a force (the restoring force) which is proportional to the distance OP from it. If x is this distance OP, we shall denote the restoring force by $-qx$, the minus sign showing that the force is in the opposite direction to the x axis. The Newtonian equation of motion is

$$m\frac{d^2x}{dt^2} = -qx,$$

of which the general solution is

$$x = A\cos(2\pi vt + \varepsilon),$$

where ε is any constant and

$$v = \frac{1}{2\pi}\sqrt{\frac{q}{m}}.$$

The potential energy of the particle is

$$V(x) = \tfrac{1}{2}qx^2,$$

so the Schrödinger equation is

$$\frac{d^2\psi}{dx^2} + \frac{8\pi^2 m}{h^2}(W - \tfrac{1}{2}qx^2)\psi = 0. \tag{4}$$

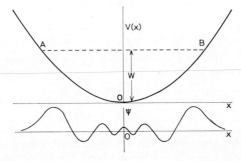

Fig. 31. Potential energy $V(x)$ and wave function ψ for a particle carrying out simple harmonic motion. The excited state illustrated has energy W equal to $(n+\tfrac{1}{2})hv$, where $n = 8$.

The potential energy and wave functions, representing the electron moving backwards and forwards, are shown in fig. 31; we have to prove that solutions of this kind can be obtained only for certain quantized values of the energy W, which we shall show to have the values

$$W_n = (n+\tfrac{1}{2})hv,$$

where

$$n = 1, 2, 3, \ldots$$

64

In Chapter 5 we gave a simple estimate of the energy values by fitting n half waves, with wavelength $\lambda = h/\sqrt{(2mW)}$, into the length $2x_0$ of the line along which the particle moves, x_0 being given by

$$\tfrac{1}{2}qx_0{}^2 = W.$$

This treatment neglected two factors.

(a) That the particle slows down as it approaches the points $x = \pm x_0$, so that the corresponding wavelength (h/mv) increases.

(b) The tunnelling of the particle beyond the points $x = \pm x_0$.

We can take the first into account by using the Wentzel–Kramers–Brillouin (WKB) approximation in a form that describes the change of wavelength with x, but not the 'tunnelling' of the particle beyond the points A and B. To use this method, we start with the Schrödinger equation (4), which we write

$$\frac{d^2\psi}{dx^2} + f(x)\psi = 0,$$

where

$$f(x) = 8\pi^2 m(W - \tfrac{1}{2}qx^2)/h^2. \tag{5}$$

The WKB solutions for *standing* waves are of the forms (cf. Chapter 6)

$$\psi = A(x) \cos\left(\int_0^x f^{1/2}\, dx\right) \tag{6}$$

and

$$\psi = A(x) \sin\left(\int_0^x f^{1/2}\, dx\right). \tag{7}$$

It is clear from symmetry that (6) gives solutions which do not change sign when x is substituted for $-x$ and so correspond to the ground state and to states with an even number of zeros; (7) gives solutions with an odd number of zeros. So solutions corresponding to standing waves will be obtained when either of the two expressions (6) and (7) vanishes at A and B, and thus if

$$\int_0^{x_0} f^{1/2}\, dx = \tfrac{1}{2}n\pi, \tag{8}$$

where $n = 1, 2, \ldots$

The integral in (8) can be evaluated by setting

$$x = (2W/q)^{1/2} \sin \theta$$

and one finds that the left-hand side of (8) is equal to

$$\frac{\pi^2 W}{h}\sqrt{\frac{m}{q}}$$

or, substituting for $\sqrt{(q/m)}$,

$$W = nh\nu.$$

65

The formula tends to the correct value for large n, but gives too high a value for the ground state ($h\nu$ instead of $\frac{1}{2}h\nu$). This comes from neglecting the tunnelling, that is the spread of the wave function beyond A and B; the half wave has been packed into too small a space.

We now describe a method obtaining an exact solution of the Schrödinger equation (4). This can be done by a standard artifice. First of all, it is desirable to reduce (4) to the form

$$\frac{d^2\psi}{d\xi^2} + (\lambda - \xi^2)\psi = 0, \tag{9}$$

by writing $x = a\xi$, $a = (4\pi^2 mq/h^2)^{1/4}$, $\lambda = 2W/h\nu$. This can be easily verified. Then we write

$$\psi = H(\xi) \exp\left(-\tfrac{1}{2}\xi^2\right),$$

which by substituting in (9) leads to

$$\frac{d^2H}{d\xi^2} - 2\xi\frac{dH}{d\xi} + (\lambda - 1)H = 0. \tag{10}$$

Now if $\lambda = 1$, $H = $ constant is a solution and this clearly corresponds to the ground state of the system, with one half wave fitted in between A and B (fig. 31). $\lambda = 1$ means $W = \frac{1}{2}h\nu$, so in the ground state the energy of a harmonic oscillator is *half* a quantum.

To find the other solutions we have to proceed as follows. The function H must be expanded in an infinite series. We write

$$H = 1 + a_1\xi + a_2\xi^2 + \ldots$$

Inserting this into equation (9), we find

$$\frac{d^2H}{d\xi^2} = \sum_n a_{n+2}(n+2)(n+1)\xi^n,$$

$$-2\xi\frac{dH}{d\xi} = -\sum_n a_n 2n\xi^n,$$

$$(\lambda - 1)H = (\lambda - 1)\sum_n a_n\xi^n.$$

If the equation is to be satisfied for all values of ξ, then the coefficient *of every power* of ξ has to vanish. Therefore, if H is a solution of the equation, for all values of n we have

$$a_{n+2}(n+2)(n+1) = a_n(2n - \lambda - 1). \tag{11}$$

This equation, then, enables us to find a_2 (since $a_0 = 1$) and thence a_4 and so on. A solution can be written down which has what we call 'even parity'; this means that it contains only even powers of x, so that $\psi(x) = \psi(-x)$. A solution with even parity is shown in fig. 31.

66

In the same way, we can construct solutions with odd parity, such that $\psi(x) = -\psi(-x)$, by writing H in the form

$$H = \xi + a_3\xi^3 + a_5\xi^5 + \ldots$$

As we have seen, a differential equation of type (9) always has two independent solutions, and this method enables two such solutions to be found, for any value of the energy W.

From equation (10) we see that if $\lambda = 2n+1$, where $n = 0$, 1, or any integer, that a_{n+2} vanishes and so do all coefficients for larger values of n. So $H(\xi)$ is a polynomial. Now the function $\exp(-\frac{1}{2}\xi^2)$ always tends to zero faster than ξ^n tends to infinity, so for these values of λ we can find at least one solution which has the desired characteristics; it oscillates in the region of ξ, where $H(\xi)$ has its zeros and for large ξ tends exponentially to zero. It follows that the values of W for which $\lambda = 2n+1$ are the quantized values of the energy for which solutions of the Schrödinger equation can be obtained which represent standing waves. Since $\lambda = 2W/hv$, this gives for the quantized energy values

$$W = (n+\tfrac{1}{2})hv.$$

It remains to be shown that if λ does *not* have one of these values, $\psi(x)$ is not a physically meaningful solution, but in fact one which tends to infinity as x becomes large, like curve (a) in fig. 26. In this case the series does not terminate, and for large n, equation (11) gives

$$a_{n+2}/a_n \simeq 2/n.$$

But this is just the way $\exp(\xi^2)$ behaves, since

$$\exp(\xi^2) = 1 + x^2 + \frac{x^4}{2!} + \ldots \frac{x^{2n}}{n!} + \ldots,$$

so that for this series

$$\frac{a_{n+2}}{a_n} = \frac{(\tfrac{1}{2}n+1)!}{(\tfrac{1}{2}n)!} = \frac{1}{\tfrac{1}{2}n+1},$$

which behaves like $2/n$ when n is large. We deduce that the series for H represents a function which goes to infinity like $\exp(\xi^2)$. Thus ψ goes to infinity like $\exp(\tfrac{1}{2}\xi^2)$. These are physically meaningless solutions, since they are not bounded, which is what we set out to show.

As explained in Chapter 2, quantum theory began with the hypothesis that certain kinds of oscillator should have quantized energy values nhv; Planck in 1900 made this hypothesis for light waves and Einstein in 1905 for the atoms in solids vibrating about their mean positions. Quantum mechanics changes these energy values to $(n+\tfrac{1}{2})hv$. Thus the atoms in a solid continue vibrating even at the absolute zero of temperature. The quantity $\tfrac{1}{2}hv$ is called the energy of the 'zero-point' motion.

In two or three dimensions the wave equation is a function of x, y or of x, y, z and the Schrödinger equation in this case has been given at the end of the previous chapter. Wave functions describing the stationary states of a particle in more than one dimension have properties that do not appear in one dimension, namely

(i) Each stationary state is defined by two quantum numbers in two dimensions and by three in three dimensions.

(ii) It may happen that two or more states with different quantum numbers have the same energy; the states are then said to be 'degenerate'. This also occurs for an electron moving on a circle, as explained in Chapter 5.

This behaviour can be shown most simply by considering a particle moving freely in a rectangular box of sides a, b. If the walls are perfectly reflecting, the wave function ψ must vanish when $x = 0$ and $x = a$, and when $y = 0$ and $y = b$. The solution of the Schrödinger equation

$$\frac{\partial^2\psi}{\partial x^2}+\frac{\partial^2\psi}{\partial y^2}+\frac{8\pi^2 m_e W}{h^2}\,\psi = 0$$

with these boundary conditions is

$$\psi = A \sin (\pi n_1 x/a) \sin (\pi n_2 y/a),$$

where n_1, n_2 are both integers, and the characteristic energy values are

$$\frac{8m_e W}{h^2} = \frac{n_1^{\,2}}{a^2}+\frac{n_2^{\,2}}{b^2}.$$

The energy is thus defined by *two* quantum numbers, n_1, n_2. The numbers $n_1 - 1$, $n_2 - 1$ give the number of nodal lines (lines on which ψ vanishes). Figure 32 shows the nodal lines for $n_1 = 4$, $n_2 = 2$.

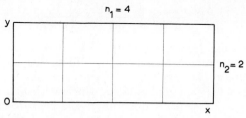

Fig. 32. Nodes, shown by thin lines, of a wave function of a particle in a rectangular box.

If $a = b$, then the state with quantum numbers n_1, n_2 has the same energy as the state with quantum numbers n_2, n_1. Thus except for states of the type (n, n)—that is with the two quantum numbers equal to each other—all states are degenerate. In this case one cannot define a unique wave function or unique pair of wave functions for the states. Consider for instance the states (2, 1) and (1, 2); they have functions (in which we have set the factor multiplying the sine function equal to 1)

$$\psi_{1,2} = \sin (2\pi x/a) \sin (\pi y/a), \tag{12}$$

$$\psi_{2,1} = \sin (\pi x/a) \sin (2\pi y/a). \tag{13}$$

But any new function of the type

$$A\psi_{1,2} + B\psi_{2,1}$$

is equally a solution; if for instance $A = B$ one gets

$$\psi = \sin \left(\frac{\pi x}{a}\right) \sin \left(\frac{\pi y}{a}\right) \cos \frac{\pi(x+y)}{2a} \sin \frac{\pi(x-y)}{2a}. \tag{14}$$

The nodes of these three functions are shown in fig. 33.

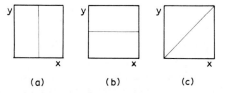

(a)   (b)   (c)

Fig. 33. Nodes, shown by thin lines, of a degenerate wave function of a particle in a square box. The first excited state is shown, and (a), (b) and (c) refer to possible degenerate states.

The physical importance of this concept of 'degeneracy' will become apparent in the next paragraph.

7.5 *Particles in a circular or spherical box*

Methods of treating these problems are of great importance, particularly for the understanding of the hydrogen atom, as will be shown in the next chapter. Here we shall discuss the solutions of the Schrödinger equation

$$\nabla^2\psi + 4\pi^2 K^2\psi = 0, \tag{15}$$

subject to the condition that ψ vanishes at the boundary of a circular or spherical box of radius a. The analysis is applicable to de Broglie waves describing a particle shut up in a circular or spherical box, to the vibration of a circular membrane and to sound waves in a spherical cavity. The full mathematical treatment of the problem requires the solution of (15) in polar or spherical polar coordinates and this will not be given in this book. The treatment that we shall give will be descriptive only.

We consider first solutions of equation (15) which are functions of r only, so that $\psi(r)$ is the same at all distances r from a given point. Solutions of this kind would describe a wave diverging from a point in space, for instance a sound wave at some distance from a point source. Any such spherical wave in three dimensions has an amplitude which falls off inversely as the distance r, and an intensity therefore which falls off as the inverse square of r; thus the amplitude of a sound wave will behave like

$$\frac{A}{r} \sin \{2\pi(Kr - vt)\}$$

or, using the complex form appropriate to quantum mechanics,

$$\psi = \frac{A}{r} \exp \{2\pi i(Kr - vt)\}. \tag{16}$$

This gives a number of particles per unit time crossing a sphere of radius r equal to

$$v|\psi|^2 4\pi r^2,$$

which is independent of r, as it should be. Similarly in two dimensions ψ falls off as $1/r^{1/2}$. A standing wave, representing a wave moving from the centre and reflected at a spherical boundary of radius a, must be compounded of the outgoing wave (16) and the reflected wave

$$\psi = \frac{A}{r} \exp \{2\pi i(-Kr - vt)\}.$$

These must be added in such a way that ψ remains everywhere finite, and we therefore *subtract* them, giving a sine function which vanishes at the origin so that the term $1/r$ does not produce a singularity. Thus ψ is of the form

$$\psi = \frac{A}{r} \sin 2\pi Kr \exp (-2\pi i vt), \tag{17}$$

where A, as before, is an arbitrary constant.

We now verify that the Schrödinger equation (15) does in fact have solutions of the forms (16) and (17). When ψ is a function of r only, it is possible (in the case of a sphere) to reduce the Schrödinger equation by some laborious calculation to the simple harmonic equation. For we may write

$$r = (x^2 + y^2 + z^2)^{1/2}$$

and hence

$$\frac{\partial \psi}{\partial x} = \frac{d\psi}{dr} \cdot \frac{\partial r}{\partial x} = \frac{d\psi}{dr} \cdot \frac{x}{r}$$

70

and therefore

$$\frac{\partial^2 \psi}{\partial x^2} = \frac{d^2\psi}{dr^2} \cdot \frac{x^2}{r^2} + \frac{d\psi}{dr}\left(\frac{1}{r} - \frac{x^2}{r^3}\right).$$

On adding $\partial^2\psi/\partial x^2$, $\partial^2\psi/\partial y^2$ and $\partial^2\psi/\partial z^2$, we find

$$\nabla^2\psi + 4\pi^2 K^2\psi = \frac{d^2\psi}{dr^2} + \frac{2}{r}\frac{d\psi}{dr} + 4\pi^2 K^2\psi \tag{18}$$

which we have to equate to zero. We now set

$$\psi = f(r)/r.$$

Then

$$\psi' = f'/r - f/r^2,$$
$$\psi'' = f''/r - 2f'/r^2 + 2f/r^3.$$

Substituting in (18) we find

$$\frac{d^2f}{dr^2} + 4\pi^2 K^2 f = 0,$$

which is the familiar simple harmonic equation of which solutions are $\cos 2\pi Kr$, $\sin 2\pi Kr$, $\exp(2\pi iKr)$. One possible solution for ψ is therefore

$$\psi = Ar^{-1}\exp(2\pi iKr), \tag{19}$$

which, as we have seen, represents an outgoing wave with the amplitude falling off inversely as the distance. The solution cannot be valid right down to the origin because one cannot have an infinite amplitude, but away from the source of the wave it should be valid. For instance, it might describe alpha particles escaping from a radioactive source and (19) would not be valid within the source. But for the problem here, there is no source of particles at the origin and such a solution is clearly not admissible, because ψ becomes infinite at the origin. $r^{-1}\cos 2\pi Kr$ is ruled out for the same reason; the only admissible solution is, as we have already seen,

$$\psi = Ar^{-1}\sin 2\pi Kr$$

which is finite at the origin.

For a particle enclosed within a spherical box of radius a, ψ must vanish when $r = a$. This imposes the quantum condition on K

$$K = n/2a,$$

where n is an integer ($n = 1, 2, 3\ldots$). The appropriate wave function is shown in fig. 34. The corresponding energy values are

$$W = h^2 n^2/8ma^2. \tag{20}$$

71

In two dimensions the solutions are somewhat more complicated, being standard Bessel functions which fall off as $r^{-1/2} \sin (2\pi K r + \text{const.})$ for large values of r. These are of little importance in atomic physics, but the two-dimensional case is of interest as illustrating solutions which are *not* functions of r only. These will now be discussed.

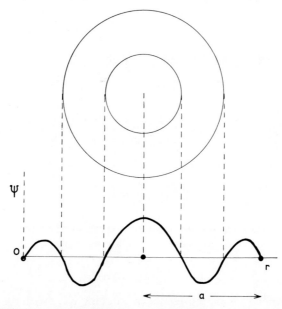

Fig. 34. Wave function ψ and nodes (shown by circles) of particle in a spherical box; the function shown is that with spherical symmetry, describing an s-state.

In quantum mechanics the states of an electron with wave functions having spherical symmetry are called 's-states'. These are by no means the only solutions of the Schrödinger equation (15). Consider first the two dimensional case and introduce polar coordinates r, θ as in fig. 35. A solution of the wave equation in polar coordinates shows that wave functions exist with lines passing through the origin on which ψ vanishes. Solutions of this kind are illustrated in fig. 36. The wave function ψ in this case can be shown to be of the form

$$\psi(r, \theta) = F(r) \cos l\theta, \qquad (21)$$

where $F(r)$ is the radial part of the wave function illustrated in fig. 36 and l is an integer; the case $l = 3$ is shown in the figure. l may take the values

$$l = 0, 1, 2, \ldots$$

$l = 0$ describes the solutions with only circular nodes. l is called the 'azimuthal' quantum number.

72

Except for the case $l = 0$, we might equally well write instead of (21)

$$\psi(r, \theta) = F(r) \sin l\theta. \qquad (22)$$

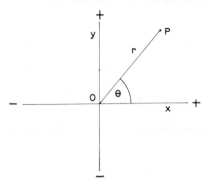

Fig. 35. Polar coordinates r, θ.

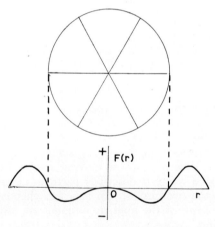

Fig. 36. Wave function and nodes for particle in spherical box, for the case $l = 3$
(an f-state).

For $l = 1$ the two cases are shown in fig. 37. The existence of two solutions means that, unless $l = 0$, the state is 'degenerate'. There are two *independent* solutions with the same energy W (or in an ordinary wave problem, such as a membrane, the same frequency). By 'independent' is meant the following. *Any* solution with this energy (or frequency) can be expressed as a sum of the two independent solutions, in the form

$$\psi = F(r)(A \cos l\theta + B \sin l\theta).$$

F

73

Such a solution (if A and B are real) would have the radial lines rotated to some position between those of fig. 37 (a) and (b).

But there is no need for A and B to be real. By putting $A = 1$ and $B = i$, one obtains a solution of the form

$$\psi = F(r) \exp (il\theta).$$

Putting in the time factor, this becomes

$$\psi = F(r) \exp [i(l\theta - 2\pi\nu t)]. \tag{23}$$

The real part is

$$\psi = F(r) \cos (l\theta - 2\pi\nu t). \tag{24}$$

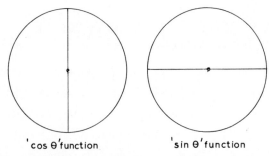

'cos θ' function   'sin θ' function

Fig. 37. Nodes of two degenerate p-functions in which the parts of the wave function depending on θ are $\cos \theta$ and $\sin \theta$ respectively.

In such a solution there are no zeros in $|\psi|^2$ passing through the origin; but if one is dealing with waves on a membrane, so that the real solution (24) is appropriate, the nodal lines *rotate* round the origin. The situation can perhaps be envisaged more easily for a wave in a medium bounded by two concentric circles—for instance a circular moat round a castle; then (23) describes waves moving round the moat. If the wave is a de Broglie wave, it describes particles circulating round in the area between the two circles. In the next chapter we shall show that l gives a measure of the angular momentum, which is $lh/2\pi$, as already outlined in Chapter 5.

The three-dimensional case is rather harder to envisage, and we shall return to this in the next chapter when the hydrogen atom is under discussion. However, the important case when $l = 1$ is simple enough. For the two-dimensional problem the wave functions are of the form $F(r) \cos \theta$, $F(r) \sin \theta$, and as $x = r \cos \theta$, $y = r \sin \theta$ we may write these $f(r)x$, $f(r)y$. In three dimensions the corresponding functions are

$$f(r)x, f(r)y, f(r)z.$$

There are *three* independent functions, corresponding to nodes on the yz, zx and xy planes.

CHAPTER 8
the Schrödinger equation for an electron in an atom

The hydrogen atom

THE hydrogen atom consists of a single electron moving in the electric field of a proton. The attractive force between the electron and the proton is $e^2/4\pi\varepsilon_0 r^2$, where e is the charge on the electron, r (in metres) the distance between them and ε_0 is the permittivity of a vacuum. The potential energy $V(r)$ of the electron when at a distance r from the proton is given, as in Chapter 1, by

$$V(r) = -e^2/4\pi\varepsilon_0 r.$$

This is plotted in fig. 38. The total energy W (kinetic and potential) of the electron is negative, if we take for the zero of energy that of an electron at rest in free space. This simply means that the energy of the electron in the atom is lower than that of the free electron at rest in free space, or that an energy $|W|$, the ionization energy, is required to remove it. The energy W is shown by the horizontal line in fig. 38. The point P shows the greatest distance that a 'classical' electron with energy W can move from the nucleus; the distance OP, which we denote by r_0, is given by

$$1/r_0 = |W|/(e^2/4\pi\varepsilon_0). \tag{1}$$

The wave description of a hydrogen atom is that of an electron shut up in a spherical box of radius r_0, as described in the last chapter. The spherical nodes and wave function ψ are shown in fig. 38. The situation is however different from that described there in three ways:

(i) The wave function does not drop to zero abruptly at the boundary but tunnels beyond it.

(ii) The kinetic energy $(W-V)$, shown in the figure, is by no means constant in the box.

(iii) The radius r_0 depends on the energy W according to equation (1).

Using (ii) and (iii) we can make a rough estimate of the quantized values of W as follows. We have seen that if the potential energy for a particle in a box of radius r_0 is constant, the wave number K is given by

$$Kr_0 = \tfrac{1}{2}n, \tag{2}$$

where $n = 1, 2, 3$, etc. In the absence of any term $V(r)$, $h^2K^2/2m_e$ would be the kinetic energy. In the hydrogen atom, as fig. 38 shows, the kinetic energy $W + e^2/4\pi\varepsilon_0 r$ varies from infinity at the origin to

75

zero at r_0. To obtain a *very* rough estimate of the energy levels, we take as an average value for the kinetic energy $h^2K^2/2m_e$ its value at $\frac{1}{2}r_0$; thus, we take as an approximation

$$h^2K^2/2m_e \simeq W + 2e^2/4\pi\varepsilon_0 r_0$$

$$= e^2/4\pi\varepsilon_0 r_0. \tag{3}$$

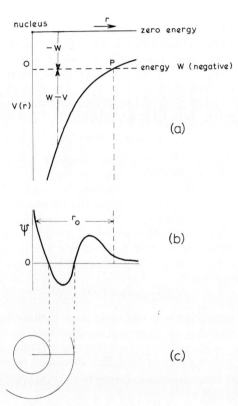

Fig. 38. (a) Potential energy $V(r)$. (b) Wave function. (c) Nodes in the wave function for an electron in a hydrogen atom.

From (2) and (3) we can eliminate K and find

$$h^2n^2/8m_e r_0^2 = e^2/4\pi\varepsilon_0 r_0,$$

which gives for the radius of the hydrogen atom

$$r_0 = (4\pi\varepsilon_0 h^2/8m_e e^2)n^2,$$

where n is an integer, $n-1$ being the number of zeros (nodes) between the origin and $r = r_0$. Apart from the numerical constant, this is just

76

the radius a_0 of the Bohr orbit obtained in Chapter 3. Also, since $-W = e^2/4\pi\varepsilon_0 r_0$, we find for the energy

$$W = -\frac{m_e e^4}{2\pi^2 \varepsilon_0^2 h^2} \frac{1}{n^2}.$$

Again, this is the correct formula, depending on $1/n^2$, apart from the numerical factor.

To obtain the correct formulae, we must use the Schrödinger equation. We consider first the spherically symmetrical solutions which we write

$$\psi = \frac{1}{r} f(r) \exp(-2\pi i\nu t), \tag{4}$$

where $f(r)$ is an oscillating function of the kind shown in fig. 38. Just as in the last chapter, we show that $f(r)$ satisfies just the same differential equation as we used in one-dimensional problems, namely

$$\frac{d^2 f}{dr^2} + \frac{8\pi^2 m_e}{h^2}(W - V)f = 0. \tag{5}$$

This can be proved quite simply by inserting (4) into the three-dimensional Schrödinger equation

$$\nabla^2\psi + \frac{8\pi^2 m_e}{h^2}(W - V)\psi = 0. \tag{6}$$

We have then to find a solution of (5) which, as in §7.5, vanishes at $r = 0$; but, instead of vanishing at r_0, it must fall exponentially to zero as r becomes larger than r_0. When r is large, (5) becomes

$$\frac{d^2 f}{dr^2} - \gamma^2 f = 0,$$

where $\gamma^2 = -8\pi^2 m_e W/h^2$. The solutions are

$$\exp(-\gamma r), \quad \exp(\gamma r)$$

and the wave function that we require must behave like the first, falling off exponentially for large values of r. This suggests that we should try to get a solution of (5) by making the substitution

$$f(r) = \exp(-\gamma r)F(r).$$

If we substitute this in (5), we find

$$f' = (F' - \gamma F)\exp(-\gamma r),$$
$$f'' = (F'' - 2\gamma F' + \gamma^2 F)\exp(-\gamma r),$$

77

so our equation for F is, if we write $\alpha - 4\pi m e^2/h^2\varepsilon_0$,

$$F'' - 2\gamma F' + \frac{\alpha}{r} F = 0. \tag{7}$$

This is an equation which can be solved in series in the same way as the equation for the harmonic oscillator. We set, since if f is to be finite, F must vanish when $r = 0$,

$$F = r(1 + a_1 r + a_2 r^2 + \ldots). \tag{8}$$

The coefficient of r^s, when we insert this in equation (7), is

   (i) from F'', $a_{s+1}(s+2)(s+1)$,
   (ii) from $-2\gamma F'$, $-2\gamma a_s(s+1)$,
   (iii) from $\alpha F/r$, αa_s.

The coefficient of r^s, when (8) is inserted in (7), must vanish identically for all s; therefore the sum of these three quantities must vanish, so that

$$a_{s+1}(s+2)(s+1) + a_s\{-2\gamma(s+1) + \alpha\} = 0. \tag{9}$$

This relationship enables us to calculate a_1 from a_0 (which is equal to 1), a_2 from a_1 and so on, and is called a 'recurrence relationship'.

The first use we make of it is to note that if for some integral value of s (including zero)

$$-2\gamma(s+1) + \alpha = 0, \tag{10}$$

the series terminates; all coefficients beyond a_s vanish and F is therefore a polynomial. Now $\exp(-\gamma r)$ always tends to zero faster than r^s goes to infinity, whatever s may be; in other words $r^s \exp(-\gamma r)$ tends to zero as r tends to infinity. Thus if (10) is valid, F vanishes at $r = 0$ and tends to zero for large r. Therefore ψ is finite at the origin and tends to zero for large r. These are the properties that we require. Thus equation (10) gives a set of energy values for which ψ has the properties of a standing wave. If we put in the values of α, γ, equation (10) gives

$$W = -W_{\rm H}/(s+1)^2, \quad s = 0, 1, 2, \ldots, \tag{11}$$

where

$$W_{\rm H} = m_e e^4/8\varepsilon_0 h^2.$$

These are exactly the values predicted by Bohr's theory of orbits. There is no obvious reason why the two theories should give the same result, and they do not for any function $V(r)$ different from the Coulomb form.

It remains to be shown that when (10) is not valid, ψ tends to infinity as $\exp(\gamma r)$. The proof is similar to that already used for the harmonic

78

oscillator. We are interested in what happens for very large r. Therefore terms in the series (8) for which s is large are the ones that matter.
For these α can be neglected in (9). Thus

$$a_{s+1}(s+2) \simeq 2\gamma a_s.$$

We note the similarity of this series to $\exp(2\gamma r)$ for which

$$a_{s+1}(s+1) = 2\gamma a_n.$$

We deduce from this that F behaves like $\exp(2\gamma r)$ and hence ψ like
$\exp(\gamma r)$. Unless (10) is valid, therefore, the solutions *diverge* at
infinity.

For the ground state the wave function is $A \exp(-\gamma_0 r)$, where $1/\gamma_0$
is the Bohr radius of the hydrogen atom, which we write a_0, so that

$$a_0 = \varepsilon_0 h^2 / \pi m_e e^2 \tag{12}$$

which is equal to

$$0\cdot053 \text{ nm} = 0\cdot53 \text{ Å}.$$

A is a constant. The interpretation of the wave function is as usual
that $|\psi|^2 \, dx \, dy \, dz$ is the probability that the electrons will be found at
any point in the volume $dx \, dy \, dz$. The most probable place for the
electron is therefore at $r = 0$, and the probability of finding it anywhere
else falls off exponentially with distance. $|\psi|^2 4\pi r^2 \, dr$ is the proba-
bility that it will be found between the distances r, $r+dr$ from the
nucleus. Since the electron must be somewhere in the atom, A must
be chosen so that

$$A^2 \int_0^\infty \exp(-2\gamma_0 r) 4\pi r^2 \, dr = 1.$$

This gives $A = 1/8\pi^{1/2}\gamma^{3/2}$.

For the other solutions, the quantum number s represents the number
of zeros in the function f, and thus the number of spheres on which ψ
vanishes. The usual notation (as in Chapter 3) is to write $n = s+1$
and to call n the principal quantum number. These solutions which
have spherical symmetry and represent an electron going backwards
and forwards between the centre and the boundary of the atom describe
states of the atom which we call s-states. The state with $(n-1)$ zeros
in ψ is called the ns-state. The s-states are states of the atom in which
the electron is not rotating round the atom, so that the angular
momentum is zero. The lowest state is always an s-state. In Bohr's
theory of orbits, states with no angular momentum did not exist, and
this is an important difference between the two theories.

These s-states however are not the only solutions of the Schrödinger
equation for the hydrogen atom; others exist which do have an angular

momentum. We have seen in Chapter 5 that a particle moving on a circle of radius R can be described by wave functions

$$\exp\{2\pi i(Kx \pm vt)\}, \tag{13}$$

where x is the distance along the circumference shown in fig. 23 and K must have the values $K = l/2\pi R$, where l is zero or an integer. We saw also that these wave functions describe an electron moving clockwise or anticlockwise round the circle with l quanta of angular momentum. But the functions (13) are not immediately applicable to a hydrogen atom, because an electron in the atom is not constrained to move at a fixed distance from the nucleus. However, if the electron is at P in fig. 35 (p. 73), the angle θ in radians is given by $\theta = x/r$, so (13) becomes $\exp\{i(l\theta \pm 2\pi vt)\}$. This suggests that solutions of the Schrödinger equation (6) may exist of the form

$$\psi = f(r) \exp\{i(l\theta \pm 2\pi vt)\}, \tag{14}$$

where $f(r)$ is the kind of function already discussed, and that they represent an electron moving backwards and forwards from the centre to the boundary *and at the same time* moving round the nucleus with angular momentum $lh/2\pi$. Such solutions have already been discussed in the last chapter.

In this book we shall describe only the s-states already discussed for which $l = 0$, and those states for which $l = 1$; the latter are called p-states. It will be noted that these p-states are 'degenerate'. This means that the two states for which the wave functions are given by (14) have exactly the same energy. Equally good solutions of the Schrödinger equation can be obtained by adding or subtracting these two solutions; thus possible wave functions are

$$\psi = f(r)[\exp\{i(\theta - 2\pi vt)\} + \exp\{i(-\theta - 2\pi vt)\}]$$
$$= 2f(r) \cos\theta \exp(-2\pi i vt) \tag{15}$$

and

$$\psi = f(r)[\exp\{i(\theta - 2\pi vt)\} - \exp\{i(-\theta - 2\pi vt)\}]$$
$$= 2if(r) \sin\theta \exp(-2\pi i vt). \tag{16}$$

These functions describe an atom in which the electron still has one quantum of angular momentum, but in which it is just as likely that the electron is going round the nucleus in one direction as in the other. The nodes of these functions (15) and (16) are illustrated in fig. 37. In the first the wave function vanishes for $x = 0$, in the second for $y = 0$.

In the foregoing argument we have used polar coordinates r, θ appropriate to motion in a plane. In three dimensions, symmetry demands that there should be a third p function for which the wave function vanishes on the plane $z = 0$. There are therefore *three* independent p wave functions, all having the same energy.

80

We have referred already to the curious 'accident' by which the quantized energy values $-W_H/n^2$ derived from the Schrödinger equation for s-states (i.e. those for which $l = 0$) are the same as those derived from the Bohr theory. Even more surprising is that, again for the Coulomb potential energy $V(r) = -e^2/4\pi\varepsilon_0 r$, the allowed energy values for solutions of type (15) satisfy the same equation. In fact, the energy of a given state is

$$-W_H/(s+l+1)^2,$$

where s is the number of zeros in the radial function $f(r)$ and $lh/2\pi$ the angular momentum. The classification of states is thus as follows. If the energies are expressed as $-W_H/n^2$ as in Chapter 3, and n is called the principal quantum number, then we have

if $n = 1$, l must be zero and there is only *one* state available;

if $n = 2$, we can have $l = 0$ giving one state, or $l = 1$ giving three states. There are thus four states in all.

Further developments of the theory for larger values of n which will not be described here give n^2 states for any value of n.

In Chapter 3, in discussing X-ray spectra, we introduced Pauli's Exclusion Principle and stated that the experimental evidence showed that $2n^2$ electrons, not n^2, can be accommodated in the states with principal quantum number n. To understand this we must introduce the electron spin and this will be done later in this chapter. First however we shall discuss how the energy levels of an atom such as lithium (a monovalent atom with atomic number $z = 3$) differ from those of hydrogen.

8.1 *Atoms of elements in the first column of the Periodic Table (the alkali metals)*

Lithium has two electrons in the inner K shell, with s wave functions of the form $A \exp(-\gamma r)$ where $1/\gamma = \frac{1}{3}a_0$. The factor $\frac{1}{3}$ arises because e^2 has to be replaced by Ze^2 with $Z = 3$. The outer electron, when in the state with the lowest energy which the Pauli principle allows, is in a 2s-state. To calculate the allowed energies we may use the Schrödinger equation (6), but we must not take the Coulomb form for the potential $V(r)$. An approximate procedure which gives quite good results and is discussed further in the next section is to take the field of the nucleus and the *average* field of the two K electrons. In other words they are treated as if they each produced a charge density $e|\psi(r)|^2$. This means that the outer electron will be attracted to the nucleus by a force $e^2/4\pi\varepsilon_0 r^2$ when it is far away, but if it ever ventures into the region occupied by the K electrons it will be pulled by the much stronger force $3e^2/4\pi\varepsilon_0 r^2$. The resulting potential energy curve is shown in fig. 39. Moreover, unlike an electron in a circular Bohr orbit, the 2s electron *does* venture in near to the nucleus. Its wave function is as shown in fig. 38 (*b*); it is just as likely to be near the nucleus as anywhere else. We should expect then that this strong field near the nucleus would lower the energy below the value for hydrogen for a state with principal quantum number 2, namely ($W_H/4 = 3\cdot3$ eV). In fact it lowers it quite a lot; the ionization energy of lithium is 5·37 eV.

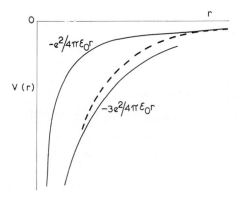

Fig. 39. Potential energy for the valence electron in the lithium atom, shown by the dotted line.

But if the electron with principal quantum number $n = 2$ is in a p-state (written 2p), it is much less likely to be in the strong field near the nucleus. Its energy is not lowered so much. So the 2s and 2p no longer have the same energies, as they do in hydrogen. Figure 40 shows the allowed energies for the outer (valence) electron of lithium and compares them with those for $n = 2$ and $n = 3$ in hydrogen.

8.2 *The electron spin*

An atom in which the electron is in a state for which the quantum number *l* does not vanish has, as well as an angular momentum $lh/2\pi$, a magnetic moment. This magnetic moment has been calculated in Chapter 5, using as our model an electron moving with velocity v in an orbit of radius r. We can consider that the atom behaves like a circular loop of wire carrying a current. The current is defined as the charge crossing any point per unit time, which is $ev/2\pi r$. The atom will behave like a magnet of moment given by

$$\mu = l\mu_0 \quad = 0, 1, 2 \ldots,$$

where μ_0 is the Bohr magneton already described in Chapter 5. The magnitude of the Bohr magneton can perhaps best be envisaged from the fact that the magnetic moment of a steel magnet can be up to between 1 and 2 times $\mu_0 N$, where N is the number of iron atoms it contains.

If an atom in a p-state (i.e. with $l = 1$) were placed in a magnetic field, we should expect the following. Suppose the magnetic field were perpendicular to the plane of the paper. Then the two wave functions described by equations (14) would no longer have the same energies. They represent states of the atom one with magnetic moment μ_0 (one Bohr magneton) parallel to the field, and the other in the opposite

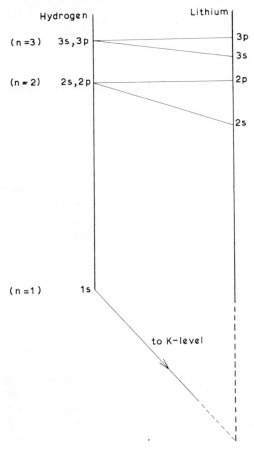

Fig. 40.   Energy levels for the valence electron in the lithium atom.

direction. If we denote the strength of the field by B, these two states have their energies shifted by $\pm \mu_0 B$; the energy level is thus split by the field. It is therefore no longer legitimate to combine the states together as in (15) and (16). The third state, however, with a node in the plane of the paper, is unaffected by the field so we expect a splitting of a p-state into three states. Such a splitting is in fact observed, and gives rise to the splitting of spectral lines when the source

83

of the radiation is placed in a magnetic field. This is known as the Zeeman effect.

However, in an s-state, and in particular in the ground state of hydrogen and of monovalent atoms, $l = 0$; therefore the theory of the motion of a point electron, as developed so far, should give no splitting in a magnetic field. It was however established in the early 1920's that in fact such levels do split; the evidence came both from spectroscopy and from experiments on beams of atoms, of which the most famous was that of Gerlach and Stern. The energy level for such a state splits into two, not three, the displacement of each level being the same as before ($\mu_0 B$).

The bold deduction was first made in 1923 by two physicists in Holland, Goudsmid and Uhlenbeck, that this could only be explained if the electron itself was not simply a point charge, but had also a magnetic moment in its own right. It behaves like a spinning sphere of charge. Though no-one now would suppose that this is what it is, this phenomenon is always called 'electron spin'. Goudsmid and Uhlenbeck ascribed to the electron one Bohr magneton of magnetic moment, but *half* a quantum of angular momentum ($h/4\pi$). They gave arguments to show that in this way the existence of a splitting into *two* states could be explained, contrasted with the splitting into three for a state with one quantum of angular momentum.

This then is the reason why, following the Pauli principle, *two* electrons can occupy every state deduced from the Schrödinger equation for point electrons such as 1s, 2s, 2p, etc. These two electrons will always have the directions of their spin moments in opposite directions, so that the helium atom, for example, with its two 1s electrons, has no magnetic moment while the hydrogen atom has a magnetic moment.

One of the most interesting effects of electron spin is that even in the absence of an external magnetic field all p-levels split into *two* levels very close together in energy, giving doublets in emission spectra such as that shown by the D-lines of sodium. If the atom is in a p-state the electron is going round the nucleus and the resulting circular current will produce a magnetic field in the atom. We estimate the magnitude of this field by supposing that the electron moves in a circular orbit of radius r. Then the field B in the middle of the orbit is given by

$$B = \mu_0 j / 2r,$$

where j, the current, is $ev/2\pi r$ as in Chapter 5, and μ_0 is the permeability of free space. Thus

$$B = \mu_0 ev / 4\pi r^2$$

and if the angular momentum $m_e v r$ has the value $h/2\pi$ appropriate to a p-state,

$$B = \frac{\mu_0 h e}{8\pi^2 m_e r^3}.$$

The energy of the electron spin in this field is $\pm B\mu_0$ which is, since $\mu_0 = eh/4\pi m_e$,

$$\mu_0 e^2 h^2 / 32\pi^3 m_e r^3. \tag{17}$$

Naturally the numerical constant has no significance, since we have taken the field B in the middle of the orbit, and it should be averaged over the whole atom. But apart from numerical constants of order unit, the energy (17) should give the magnitude of the splitting.

Of particular interest is the ratio of (17) to the ionization energy of an atom. If we take the latter for hydrogen, and insert formula (12) for the Bohr radius, we find, since $\varepsilon_0 \mu_0 = 1/c^2$,

$$\frac{\text{spin–orbit energy}}{\text{ground state energy}} = \alpha^2,$$

where

$$\alpha = \frac{e^2}{4\pi\varepsilon_0} \cdot \frac{2\pi}{hc}.$$

This term α has the dimensions of a pure number. It can alternatively be expressed as (v/c), where v is the velocity defined above; v/c is the ratio of the velocity v of the electron in the Bohr atom to the velocity c of light.

This constant has approximately the value $1/137$. The fact that it is small is of great importance for physics; were it not so, all velocities of electrons in atoms would be comparable with that of light, and no non-relativistic form of quantum mechanics would be possible. The effects due to spin, such as the splitting of the p-states, would no longer be a small correction but a major one. But we have as yet no theory which tells us why this constant should have the value it does.

Almost as soon as quantum mechanics was discovered physicists attempted to incorporate the spin in the mathematical formalism. The decisive step was however taken by Dirac† in 1928, who showed that a satisfactory relativistic wave equation could not exist for a point electron, but that the simplest possible such equation described an electron with all the spin properties which an electron is observed to have.

---

† P. A. M. Dirac, *Proc. R. Soc.* A, **117**, 610, 1928.

CHAPTER 9

the many-electron atom

AN atom with atomic number Z consists of Z electrons moving in the field of the nucleus which carries a charge Ze; the atoms known in nature have values of Z from unity for hydrogen up to 92 for uranium, with higher values for the artificial transuranic elements. The purpose of this chapter is to give a description in terms of quantum mechanics of the properties of atoms containing more than one electron.

To a certain approximation it is sufficient to treat each electron as moving independently of all the others, and having its own wave function $\psi(x, y, z)$ and its own energy. This is the treatment that will be adopted in this chapter up to the last section. But it must be strongly emphasized that this is an approximate treatment, because it neglects, or rather averages, the potential energy $e^2/4\pi\varepsilon_0 r$ of a pair of electrons at a distance r from each other. Further developments of quantum mechanics make it possible to treat all the Z electrons together as one dynamical system. Nevertheless the approximation of independent electrons is extremely useful, and we should hardly be able to understand, say, the Periodic Table or X-ray spectra without it.

If each electron is to have its own wave function, we have to ask what potential energy $V(r)$ we ought to insert in Schrödinger's equation. It was first proposed by the English physicist Hartree† in 1928 that the appropriate potential energy for each electron would be that $(-Ze^2/4\pi\varepsilon_0 r)$ due to the nucleus together with the *averaged* potential energy due to all the other electrons. This is calculated in the following way. Consider the contribution to the potential energy $V(r)$ of a given electron made by another electron which has wave function $\psi(x, y, z)$. The chance that this electron is in the volume element $dx\, dy\, dz$ is $|\psi(x, y, z)|^2\, dx\, dy\, dz$. Thus, *on the average* it will produce a charge density

$$e|\psi(x, y, z)|^2.$$

The type of charge density envisaged is shown in fig. 41. The assumption made by Hartree is that the field due to this charge, giving a contribution to the potential energy as shown in fig. 39, acts on all the other electrons. Consider the application of these ideas to the helium atom, which contains two electrons both with the same wave function $\psi(r)$. The argument is similar to that given for lithium in the last chapter. The potential energy $V(r)$ must behave like $-2e^2/4\pi\varepsilon_0 r$ for small values

† D. R. Hartree, *Proc. Camb. Phil. Soc.*, **24**, 89, 1928.

of r and like $-e^2/4\pi\varepsilon_0 r$ when r is large. For intermediate values it cannot be calculated unless $\psi(r)$ is known, and conversely $\psi(r)$ cannot be calculated until V is known. The wave functions and potential energies are found, therefore, by a method of trial and error and are called 'self-consistent'. In an atom with atomic number Z, the function $V(r)$ behaves like $-Ze^2/4\pi\varepsilon_0 r$ near the nucleus, and like $-e^2/4\pi\varepsilon_0 r$ far from it.

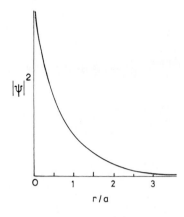

Fig. 41. Density of charge $|\psi|^2$ within an atom.

The concept of the self-consistent potential will now be used to give a description of the periodic structure of the elements, their X-ray spectra and the optical spectra of some of the simplest of them. These have to be understood in terms of the important concept we introduced in Chapter 3 in our discussion of the Bohr atom, namely Pauli's Exclusion Principle. This was introduced by the Austrian physicist Pauli in 1925 as an empirical law to explain the known facts; in the quantum theory of many-particle systems it appears as a special case of a more general law which will be briefly explained at the end of this chapter. The empirical law states: "In an atom or other system with quantized energy levels, not more than one electron can ever be in the same quantum state."

Let us look at the Periodic Table in terms of this law. The hydrogen atom has one electron in a 1s-state; the electron has no orbital angular momentum (i.e. no angular momentum due to its motion), but the electron's spin gives it both angular momentum and a magnetic moment. The electron's angular momentum is $\frac{1}{2}h/2\pi$, giving *two* possible directions in a magnetic field according to the direction of the moment. The helium atom, which has two electrons, can accommodate them both in 1s-states, but by Pauli's principle the spins must be in different states. If one imagines a magnetic field laid on to give an axis of

87

reference, one electron will have its spin parallel and the other anti-parallel. A helium atom in its ground state therefore has no magnetic moment. Its ground state is not degenerate, and its quantized energy

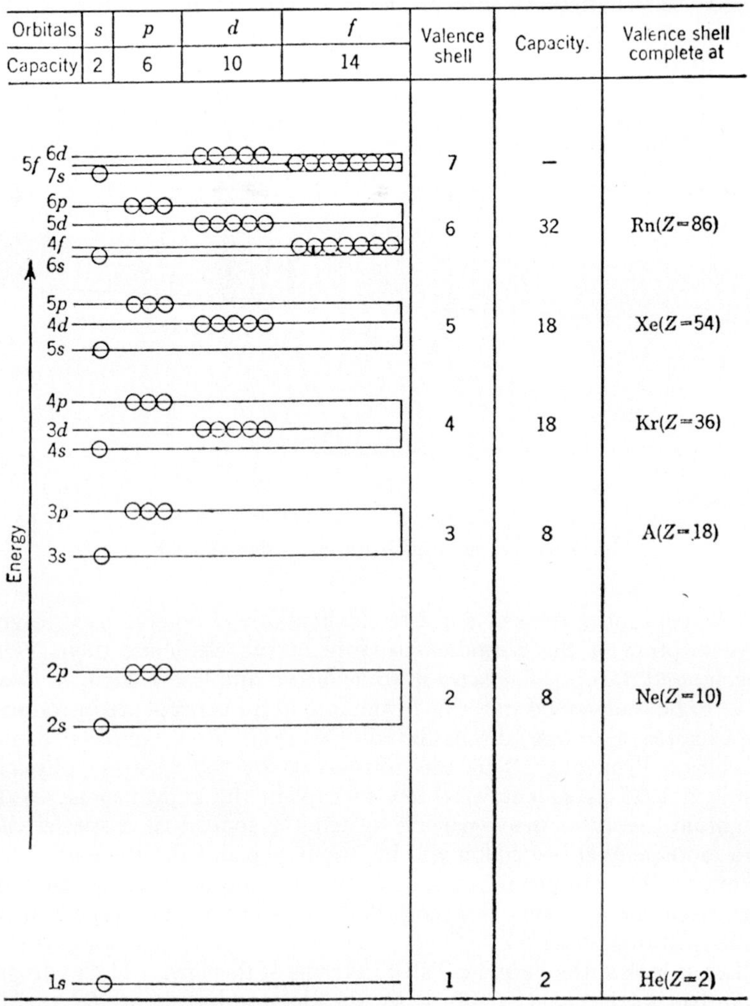

Orbitals	s	p	d	f	Valence shell	Capacity.	Valence shell complete at
Capacity	2	6	10	14			

Energy level diagram (reading from top):

Orbitals	Valence shell	Capacity.	Valence shell complete at
5f 6d 7s	7	–	
6p 5d 4f 6s	6	32	Rn($Z=86$)
5p 4d 5s	5	18	Xe($Z=54$)
4p 3d 4s	4	18	Kr($Z=36$)
3p 3s	3	8	A($Z=18$)
2p 2s	2	8	Ne($Z=10$)
1s	1	2	He($Z=2$)

Fig. 42   Reproduced from T. Moeller, *Inorganic Chemistry*, p. 94, John Wiley, 1952.

does not split in two in a magnetic field. The configuration of electrons in a helium atom is called a 'closed shell', and it will be shown in Chapter 10 why such atoms are chemically inert.

We turn now to lithium with three electrons. The third electron cannot go into the 1s-state; it must be in a 2s-state. The two electrons

88

in 1s-states will have the quantized energy values given roughly by

$$Z^2 m_e e^4 / 8 \varepsilon_0{}^2 h^2$$

and with $Z = 3$; this gives us already quite a large value, about $2\cdot4 \times 10^{-12}$ J or 150 eV; moreover the radial extension is small; the wave function is $C \exp(-r/a)$, where a is now

$$a_0 = \varepsilon_0 h^2 / \pi m_e e^2 Z = 0\cdot53 \text{ Å}/Z$$

$$= 0\cdot18 \text{ Å}.$$

These inner electrons are very little affected by the chemical state of the atom even for lithium, and very little indeed for elements farther up the Periodic Table; they are responsible, as we shall see, for X-ray spectra.

The third electron is responsible for the chemical properties and optical spectra of lithium. As we have seen, it is in a 2s-state. The ionization energy will however be greater than the value given by the Bohr formula $R_\mathrm{H}/4 = 3\cdot3$ eV, because the potential energy $V(r)$ for small r is (numerically) greater than $e^2/4\pi\varepsilon_0 r$. Actually the ionization energy of lithium is $5\cdot4$ eV. The first excited state of lithium is a 2p-state. Since p-wave functions have a zero at the origin, the energy is much less affected by the large values near the nucleus of $-V(r)$, and is close to the hydrogen values. The energy levels for lithium were shown in Chapter 8.

The lithium atom in its ground state has a magnetic moment due to the 'spin' moment of the electron, and this is true of the excited s-states too; the energy of any of the states splits into two in a magnetic field. The p-states of lithium (and of the other alkali metals) are however observed to be doublets; the well-known doublet shown by the 'D-line' of sodium is an example. This doublet is due to the electron's spin. Since in a p-state, unlike in an s-state, the electron is moving round the nucleus and has angular momentum, it sets up a magnetic field within the atom, which leads to two energy states for the electron's spin. A description of this behaviour was given in the previous chapter.

We now ask, as one goes up the Periodic Table from lithium, through beryllium, boron and carbon onwards, how many electrons can go into states with principal quantum number 2. The answer is clearly eight; two can go into the 2s-state and six into the 2p-states, since a p-state is threefold degenerate. This brings us to neon, a noble gas, with $Z = 10$; the ten electrons form closed shells with no magnetic moment. The absence of the magnetic moment has its origin in the fact that, for each electron with a wave function describing an electron going round in one direction, there will be another describing an electron going round in the opposite direction, and also for any electron with its spin parallel to a magnetic field there will be another with its spin in the opposite direction.

G

89

If the principal quantum number n is 3, the angular momentum quantum number l can have the values 0, 1 or 2. The number of available states is therefore

number of functions

$l = 0$	1
$l = 1$	3
$l = 2$	5

giving a total of 9. Since for each state there are two spin directions, the closed shell for $n = 3$ contains 18 electrons.

At first sight one might expect an atom with

$$2 + 8 + 18 = 28$$

electrons, filling up the shells with quantum numbers 1, 2 and 3, to be a noble gas like neon or argon. Actually the element nickel has 28 electrons, and when not excited has only 8 instead of 10 electrons in the state for which $n = 3$ and $l = 2$ (known as the 3d-state), and two electrons in the 4s-state ($n = 4$, $l = 0$). Nickel is what is known as a transition metal, the 3d-shell not being fully occupied. It is among the transition metals that the ferromagnetic metals iron, nickel and cobalt are found. The reason is believed to be the following. The 3d-state, not being fully occupied, can and does have a magnetic moment; and the radial extent of the 3d wave functions being smaller than those of the 4s-states occupied by the outer electrons, the moments within the atoms are not greatly affected by the cohesive forces in the solid. Many transition metals, therefore, are strongly paramagnetic and in some, for reasons which are not yet entirely clear, the moments interact in such a way as to give ferromagnetism. Figure 42 (a) shows the sequence in which the electron shells build up.

9.1 *X-ray spectra*

X-rays are normally generated by the impact of fairly energetic electrons (accelerated under several kilovolts) on a metal target. The radiation emitted from the target consists of a continuous spectrum on which are superimposed a number of sharp lines (cf. fig. 8). The continuous background has already been mentioned; the electron with initial energy $\frac{1}{2}m_e v^2$ can, on passing through the strong field within the nucleus, emit a quantum of radiation with any frequency up to v_0, where $hv_0 = \frac{1}{2}m_e v^2$. The wavelengths of the sharp lines, however, depend on the particular metal of which the target is composed, and depend little on its state of chemical composition. Thus if the surface of a copper anode is oxidized, this does not affect appreciably the wavelength of the copper lines.

The characteristic X-ray lines are due to transitions between the lower levels in an atom, which are normally occupied by as many electrons as the Pauli Principle allows. Take the case of copper, in which the states with principal quantum number $n = 1$ contain 2 electrons, those with $n = 2$ contain 8 and those with $n = 3$ contain 18. The atomic number is 29, so there is one electron which in metallic copper is free. The levels have already been shown in fig. 12 of Chapter 3, with the usual notation according to which the level for which $n = 1$ is called the K-level, that for which $n = 2$ the L-level and so on. If now in the target of an X-ray tube one of the electrons in the K-level is hit by an electron with sufficient energy and knocked out of the atom, then an electron in the L or M-level can make a transition to the 'vacant place' in the K-level, emitting an X-ray in the process. The line when an electron makes the transition from the L-level is called the Kα line, from the M-level the Kβ line.

The frequencies of these lines are given by fairly simple formulae. To the first approximation the energy of an electron in the K-level will be $Z^2 W_H$ but this will not be quite correct†; each electron in the K-shell is slightly screened from the nucleus by the other electron, so it is better to write

$$W_K = -(Z-\sigma)^2 W_H,$$

where $\sigma \sim \frac{1}{2}$. Similarly for the L-level

$$W_L = -(Z-\sigma')^2 W_H / 4,$$

where σ' will be between 2 and 10. The frequency of the Kα line is given by

$$h\nu = W_L - W_K.$$

It will be seen from these formulae that the frequency of the Kα line increases steadily with atomic number. In the early days of X-ray diffraction, when the Bragg law for reflection of X-rays from crystals had just been discovered and it was therefore possible for the first time to measure the wavelengths of X-rays, Moseley, in the University of Manchester, selecting a particular characteristic line, the Kα line, plotted $\nu^{1/2}$ (where ν is the frequency) against the atomic number Z (which until round about this time was thought of only as the position of an element in the Periodic Table for a number of elements). The variation of $\nu^{1/2}$ with Z was linear; this continuous increase of frequency with atomic number was very striking and clearly had to correspond to the increase of something more than just a 'serial number'. Moseley, just before the Rutherford scattering experiment which gave the most direct evidence for this, guessed that this something must be the number of electrons in the atom.

† W_H is here the ionization energy of hydrogen.

91

9.2 Quantum mechanical explanation of the Exclusion Principle

This lies outside the parts of quantum mechanics which are developed in this book, and this section is intended as only a very brief introduction. We have described a system with two electrons, such as the helium atom, by giving each electron its own wave function; if (x_1, y_1, z_1) are the coordinates of one electron, (x_2, y_2, z_2) those of the other, we have supposed that one wave function, which we denote by $\psi_a(x_1, y_1, z_1)$, will describe the position of one of them, and $\psi_b(x_2, y_2, z_2)$ that of the other. But if the interaction between the electrons were included in our analysis properly, we should not be able to assume the probability that the first electron is at some point in the atom to be independent of the position of the second electron; for instance, it is unlikely that both electrons will be at the same point, when the Coulomb repulsion $e^2/4\pi\varepsilon_0 r$ between them becomes infinite. Quantum mechanics in fact demands that we use a *two-electron wave function* which depends on the positions of both coordinates. If we write q_1 for $(x_1 y_1 z_1)$, and q_2 for $(x_2 y_2 z_2)$, this function can be written $\Psi(q_1, q_2)$. The interpretation of this function is that

$$|\Psi(q_1, q_2)|^2 \, dx_1 \, dy_1 \, dz_1, \, dx_2 \, dy_2 \, dz_2 \tag{1}$$

should denote the probability that one electron is in the volume element $dx_1 \, dy_1 \, dz_1$ at the point q_1, and the other particle in the volume element $dx_2 \, dy_2 \, dz_2$ at the point q_2. Quantum mechanics provides a generalization of the Schrödinger equation for these wave functions describing two (or more) particles. However it is not necessary to write down this equation in order to understand the Exclusion Principle. The essential point is that the function $|\Psi|^2$ must, to make sense, be symmetrical in the two coordinates; this means that

$$|\Psi(q_1, q_2)|^2 = |\Psi(q_2, q_1)|^2. \tag{2}$$

The quantity (1) gives the probability that, if a conceptual experiment is carried out to locate an electron, one will be found at q_1 and another at q_2. One cannot distinguish two electrons from each other; they are as alike as two pins—indeed much more so; and we ought not to ask for the chance that electron number 1, distinguished from the other by a dab of paint, is at q_1 and the other at q_2. To mark an electron with a dab of paint, or in any other way, is an experiment that can never be carried out, and all we can ask for is the chance of finding volume elements at points q_1 and q_2 each occupied by an electron.

Accepting this, then, let us return to the model used in the rest of this chapter in which each electron has a separate wave function. The two-electron functions might be

$$\psi_a(q_1)\psi_b(q_2) \tag{3}$$

which could be interpreted in the same way as the quantity (1), namely that

$$|\psi_a(q_1)|^2 \, |\psi_b(q_2)|^2 \, dx_1 \, dy_1 \, dz_1 \quad dx_2 \, dy_2 \, dz_2. \tag{4}$$

is equal to the chance that electron (1) is at q_1 and electron (2) at q_2. But this function (4) does not have property (2). We have to set up a new function that does. To do this, we note that the function (3) is degenerate; it has exactly the same energy as the function obtained by exchanging q_1 and q_2, namely

$$\psi_a(q_2)\psi_b(q_1).$$

Therefore, as in so many other places in this book, we know that combinations of the type

$$A\psi_a(q_1)\psi_b(q_2) + B\psi_a(q_2)\psi_b(q_1)$$

are also solutions of the Schrödinger equation. The combinations which give $|\Psi|^2$ the desired property (2) are

$$\psi_a(q_1)\psi_b(q_2) + \psi_a(q_2)\psi_b(q_1), \tag{5}$$

$$\psi_a(q_1)\psi_b(q_2) - \psi_a(q_2)\psi_b(q_1). \tag{6}$$

The first is called the symmetrical solution, the second the anti-symmetrical.

The question is, which ought one to take? There must be a law of nature which tells us which to take, if we are to be able to use quantum mechanics to make any prediction at all about systems of two or more particles. There is a host of evidence that for electrons (and indeed for protons and neutrons) one ought to take the antisymmetrical solution. The strongest evidence is the Exclusion Principle itself. If ψ_a and ψ_b are the *same* functions, which means that the two electrons are in the same state, it is clear that the antisymmetrical function (6) vanishes. This means that no antisymmetrical wave function is possible for two particles in the same state; in other words, if quantum mechanics describes Nature correctly, two particles cannot be in the same state.

As we have seen, the Exclusion Principle is valid when the 'state' is defined as describing not only the position of the electron but also in which of the two possible states the spin finds itself. Thus the equations (1), (5) and (6) as we have described them are not quite correct; q should refer not only to the positional coordinates (xyz) of the electron but also to some coordinate describing the spin. In this book we have not introduced this coordinate, so we shall not define q further. The point however remains that, since Nature for reasons unknown to us tells us to choose the antisymmetrical state, Pauli's Principle follows.

All the fundamental particles of physics and others that are not fundamental such as complex nuclei or atoms have the property that a wave function describing a pair of them must be either symmetrical or antisymmetrical; the former are called 'bosons' and the latter 'fermions', after the physicists Bose and Fermi who first discussed the application of statistical mechanics to particles with these properties.

CHAPTER 10

molecules and solids

10.1 *The chemical bond*

THE first molecule to which quantum mechanics was applied† was that of hydrogen (H_2). The two hydrogen atoms form what is called a homopolar bond holding them together. The word homopolar means that there is no transfer of charge from one atom to the other.

The way a chemical bond such as that in the molecule H_2 is treated in quantum mechanics is the following. One seeks first of all the solution of a Schrödinger equation for the electrons in the field of both nuclei and thus with the potential energy illustrated in fig. 43 (*a*).

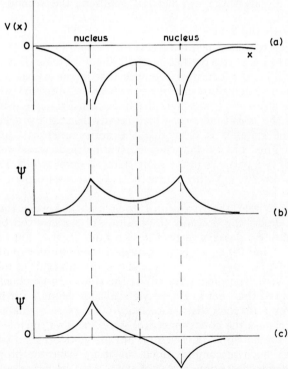

Fig. 43. (*a*) The potential energy $V(x)$ of an electron in the hydrogen molecule. (*b*) The wave function of the electron for ground state. (*c*) The wave function for the first excited state.

† By F. London and W. Heitler in 1928.

94

Just as for an atom, a solution exists with an energy level W corresponding to the ground state of the molecule. The value of W depends on the distance R between the nuclei; we shall write it $W(R)$. The two nuclei are then treated as if they were coupled together by a force of which the potential energy is

$$e^2/4\pi\varepsilon_0 R + W(R).$$

The first term is the potential energy resulting from the Coulomb repulsion between the nuclei. The behaviour of the two terms is shown in fig. 44, as is the equilibrium distance R_0 between the molecules, and the binding energy, or heat of formation W_0, liberated

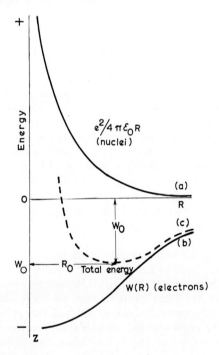

Fig. 44. The energy of H_2 as a function of the distance R between the nuclei. (a) The energy $e^2/4\pi\varepsilon_0 R$ of the nuclei in each other's field. (b) The energy of the electrons. (c) The total energy.

when the molecule is formed. The form of the wave function for the lowest state is shown in fig. 43 (b). Just as for the lowest state of an s-state in an atom, two electrons can be accommodated in the lowest state of the H_2 molecule, the directions of the spin moments being opposite to each other, so that the molecule has no magnetic moment.

95

The reasons why the energy of the electron decreases when the atoms approach each other are the following:

(*a*) As we have seen (Chapter 5), the kinetic energy of an electron confined in a box of length a is $h^2/8m_ea^2$. When the atoms come near enough together for the electrons to tunnel very easily from one atom to the other, they find themselves in a larger space than they would in either atom. This increases the value of a that one should take and so lowers the kinetic energy.

(*b*) The potential energy $V(x, y, z)$ in the region between the atoms in fig. 43 is lower than it would be in an isolated atom. This means that the energy W is lowered due to this cause.

The point Z in fig. 44 can be calculated from elementary considerations. When the atoms are a long way apart, the energy of the two electrons is $2W_H$, where $-W_H$ is the ionization energy of hydrogen, which gives

$$2W_H = -2 \times 13 \cdot 4 = 26 \cdot 8 \text{ eV.}$$

When $R = 0$, on the other hand, the energy is the same as that of two electrons in the helium atom. The ionization energy of helium is 24 eV and the energy to remove the second electron is $4W_H$ or 53 eV. There is a very large drop of about 50 eV, therefore, and the comparatively smallness of the binding energy, about 4·3 eV, arises because the term $e^2/4\pi\varepsilon_0R$ compensates for the negative energy of the electron.

An exact calculation of $W(R)$ is only possible for the very simplest case, the hydrogen molecular ion $H_2{}^+$; for all other molecules approximate methods of varying degrees of accuracy have to be used. None the less, the computer based calculations of modern theoretical chemistry give quite an accurate description of the wave functions of many molecules.

The electrons in the hydrogen molecule have excited states, and the form of the wave function for the lowest excited state for an electron in this molecule is shown in fig. 43. The energies of these states, and the frequencies of the lines emitted, bear no resemblance to those of the free atom. Moreover, to the quantized energy of the electrons must be added the energy of the molecule as it rotates about its centre of mass and the energy of the nuclei when they vibrate about their mean distance R.

Rotation of a molecule has already been mentioned in Chapter 1. The angular momentum of the rotation is quantized. If the angular velocity with which the molecule rotates about its centre of mass is ω, the kinetic energy is

$$\tfrac{1}{2}I\omega^2,$$

where I is the moment of inertia about this axis equal to $2M(\tfrac{1}{2}R)^2$. The angular momentum is

$$I\omega.$$

96

If this is quantized and set equal to $lh/2\pi$, one finds for the energy W_l

$$W_l = l^2 h^2 / 8\pi^2 I.$$

This is actually not quite correct. It would be correct if the molecule was constrained to rotate in a plane, but the extra degree of freedom which the three-dimensional problem allows leads, when the energy is evaluated by the use of the Schrödinger equation, to the value

$$W_l = l(l+1)h^2 / 8\pi^2 I. \tag{1}$$

This will not be proved here. The most important thing to realize about equation (1) is that, because the large mass M occurs in I and thus in the denominator, the energy levels are closer together by a factor of order m_e/M than the energy levels of the electron. This factor m_e/M is $1/1840$ if M is the mass of the hydrogen atom and even smaller for other molecules. This means that the energy levels of the molecule as a whole are very close together. As shown in fig. 45, for each *electronic* level there is a large number of rotational levels and observations of the separate lines necessitate spectroscopes of high resolving power, without which the absorption spectra appears as broad bands. Molecular spectra, therefore, are referred to as 'band spectra', in contrast to the line spectra emitted by atoms.

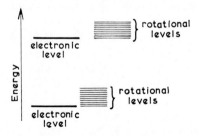

Fig. 45. Showing the electronic and rotational energies of a molecule.

10.2 *The water molecule*

As an illustration of a more complicated molecule we shall discuss the case of water, H_2O. Electron diffraction and X-ray diffraction have established the structure of this molecule which is illustrated in fig. 46 (*a*) and (*b*); the asymmetrical form leads to a preponderance of negative charge on the oxygen side of the molecule, the hydrogen atoms carrying on the average less than one electron each, so that the molecule as a whole has an electrical dipole moment. The fact that the water molecule has a dipole moment is of major importance for chemistry and biology. It is because of this dipole, for instance, that water can dissolve many minerals, such as common salt (NaCl). This is a crystalline material in which the ions Na^+ and Cl^- are held together by the strong electrostatic force between them. Salt would not dissolve in water if the ions did not adhere to the surrounding water molecules with nearly as much energy as the ions in crystalline NaCl adhere to

(a)

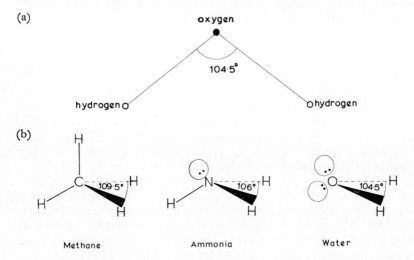

(b)

Methane   Ammonia   Water

Fig. 46. Showing positions of the nuclei in a water molecule (H_2O). and molecules of methane, ammonia and water, with the bond angles.

each other. They do this only because H_2O has a dipole moment; the water molecules round, for instance, a sodium ion in water forms what is called a hydrated shell, the molecules near to it having their negative charge turned towards it as in fig. 47.

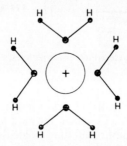

Fig. 47. Hydrated shell of water molecules round a positive ion.

We have to ask, then, why the water molecule has the form shown in fig. 46. The atomic number Z of oxygen is eight and this means that of the eight available electrons in the oxygen atom two are in the K shell with 1s-wave functions and six in states with principal quantum number $n = 2$. Of these two can go in the 2s-state, leaving four to be distributed among the six available 2p-states. Now a 2p-wave function, already illustrated in fig. 36, is as shown in fig. 48. Figure 48 (a) shows the region where the electron is, fig. 48 (b) shows the wave function. If a hydrogen atom is added at a point in the direction in

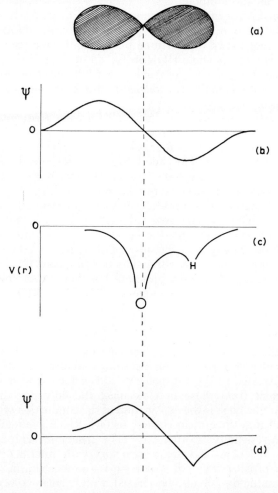

Fig. 48. (a) Region where probability of finding an electron is high in a p-state. (b) A p-wave function. (c) The potential energy of an electron in a water molecule. (d) The p-wave function distorted by the field of the proton.

which the 2p-function sticks out, the potential energy becomes as in fig. 48 (c) and the wave function takes the form of fig. 48 (d). If then the p-orbital contains only one electron, that electron and the electron from the hydrogen atom can between them occupy the state which has the wave function illustrated in fig. 48 (*d*) and form a bond similar to that in H_2, containing two electrons with antiparallel spin. It is not however completely homopolar because more charge remains on the oxygen than on the hydrogen atom, as we know from the fact that the water molecule has a dipole.

The other hydrogen atom must attach itself to an electron of which the wave function is one of the other p-functions. Since the three p-wave functions stick out in three perpendicular directions, our first conclusion is that the angle HOH in fig. 46 ought to be a right angle. The fact that it is greater is probably due to the positive charge on the hydrogen atom, incompletely screened by the electrons, which leads to some repulsion between them.

An alternative approach to the problem, one frequently used in chemistry textbooks, is as follows. Consider three common molecules in which the central atom has four pairs of electrons around it: methane, ammonia, and water. The bond angles, determined by electron diffraction and other methods, are as shown in fig. 48 (*a*). The methane molecule is rigorously tetrahedral. This is as expected from simple consideration of the mutual repulsion of electron pairs. Pairs of electrons which are not involved in bonding will, with no other nucleus to influence them, tend to be 'closer' to the central atom compared with bonding pairs. Because of this, non-bonding pairs will repel bonding pairs more forcefully than bonding pairs repel each other. This in turn causes the bond angles to be less than the tetrahedral value. This is clearly observed in the sequence CH_4, NH_3, H_2O.

10.3 *Electrons in solids*

The methods used to describe the behaviour of electrons in solids are very similar to those used for electrons in molecules, the crystalline solid being treated as a 'giant molecule'. But for solids new problems arise, different from those in molecules; the only one that will be described in this book is the explanation of the conduction of electricity.

Solids fall into three main classes; metals, which conduct electricity well, with a resistivity of order 10^{-3} Ω m, insulators with a resistivity of perhaps 10^{10} Ω m, and semiconductors with intermediate values, typically of order 1 Ω m. But as regards the mechanism of conduction semiconductors should be classed with insulators because they show the same behaviour at low temperatures; as the temperature is lowered, the resistance gets bigger and bigger, while for metals the reverse is the case, the resistance becoming smaller at low temperature. We deduce that in metals the atoms have lost at any rate their outer electrons and that these electrons are free to move in the metal and are not stuck to individual atoms, while in insulators and semiconductors the electrons are stuck to the atoms and can only move if they are shaken free in some way by the thermal vibration of the atoms.

Suppose then we take a one-dimensional model of a metal and consider a row of sodium or silver atoms; these atoms have one electron outside a closed shell and we shall call them one-electron atoms. The quantum-mechanical treatment of a metal is entirely similar to that of a hydrogen model. For simplicity let us consider a one-dimensional model, namely a row of one-electron atoms; the potential in which each electron will move is shown in fig. 49 (a). To avoid drawing 3s-functions which would be correct for sodium, we suppose these are hydrogen atoms. Then the wave function with the lowest energy will

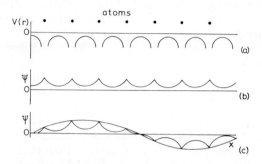

Fig. 49. (a) Potential energy of an electron in a crystal. (b) The wave function of an electron in the lowest state in the crystal. (c) The wave function of an electron moving through the crystal.

be just like that for H_2 but extended through the lattice as shown in fig. 49 (b). There are also states which represent a wave moving through the lattice with some velocity v and corresponding wavelength λ; such a function is shown in fig. 49 (c). Because the electron is moving it has higher energy than that of the state represented by the wave function of fig. 49 (b). In order to imagine the electron as carrying a current, we suppose the array of atoms to be bent round into a ring of circumference L as in Chapter 5; then λ can have the values given by the equation

$$n\lambda = L,$$

where n is zero or an integer. But n cannot have a value greater than

$$n = \tfrac{1}{2}N,$$

where N is the number of atoms; if it has this value, the wavelength is twice the distance between atoms, and as long as the wave function is made up of atomic s-functions it cannot be shorter than this.

The energies of these states are spread over a range of values; in the jargon of the subject we speak of a 'band of energies'; just as for a free electron, the faster an electron is moving the higher is its energy. Now for each value of n, except 0 and $\tfrac{1}{2}N$, *two* solutions of the Schrödinger equation exist, one representing a wave going round the

101

circle clockwise and one anticlockwise. There are therefore exactly N states of the kind illustrated in fig. 49. Since, according to the Pauli Principle, each state can accommodate *two* electrons, we see that in a solid of one-electron atoms half the states in the band will be occupied, while in a solid which has two electrons per atom all the states are occupied; the band is said to be full.

It was first pointed out by A. H. Wilson in 1931 that the electrons in states lying in a fully occupied band cannot carry a current. Just the same number of electrons are moving round the circuit in one direction as in the other, and the application of an electromotive force round the circuit will not change this. A very remarkable description of insulators was therefore proposed; one uses wave functions of characteristic free electrons, as illustrated in fig. 49; in the model any one electron can move but for the whole assembly there can be no current.

If the band is half filled, one can have more electrons moving in one direction than in the other and if a field is applied this is what happens and a current is set up. One-electron atoms invariably behave as metals in the solid state (with the exception of hydrogen which is a molecular crystal), and one supposes that a band of energy levels is formed from the state of the outer electrons in the atom (3s in sodium, 4s in potassium or copper), and that this band is half full. The energies are extended over a range of several electron volts up to a limiting

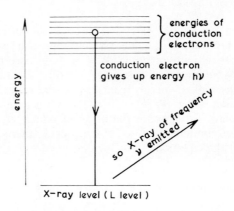

Fig. 50. An electron from the conduction band of a metal falls to fill a vacancy in an X-ray level with the emission of X-radiation.

energy (called the Fermi energy). One of the most direct ways of showing experimentally that this is so is by observing the X-ray emission due to an electron falling from one of these levels to an inner X-ray level; the process is illustrated in fig. 50. A target of, for instance, sodium is bombarded with electrons so that L electrons are knocked out from the sodium atom, and an X-ray quantum is emitted by one of the free electrons falling down to fill up the vacant place. An X-ray *band* is observed, and the shape is shown in fig. 51; the range of energies extended over 13 eV, and has a sharp upper limit.

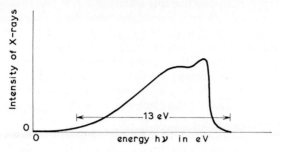

Fig. 51. The L_3 emission spectrum of aluminium.

In an insulator or semiconductor it is supposed that two (or more) bands of allowed energy levels exist and that one (the valence band) is full and the other (the conduction band) is empty (fig. 52). It is possible in principle for two such bands to overlap; this is what happens in metals with two electrons outside a closed shell, like calcium or zinc.

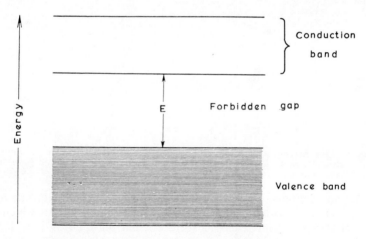

Fig. 52. The conduction and valence bands of an insulator or semiconductor. E is the width of the forbidden gap.

In an insulator they must not overlap. There must be an energy gap, denoted by E in fig. 52, separating them. This quantity is called the 'band gap'. Then a current can occur only if the energy of an electron is lifted either by thermal excitation or by the absorption of light from the valence band into the empty conduction band. At low temperatures and in the dark, the number of such electrons will become very

103

small, and the conductivity will tend to zero as the temperature is lowered. One kind of semiconductor ('intrinsic') is a material for which E is small enough for some mobile electrons to be produced at room temperature; in practice this means that E is less than about 1 eV.

The elements silicon and germanium used for transistors and rectifiers, however, owe their conductivity mainly to the deliberate introduction into the material of very small quantities of some impurity, perhaps one part in 100 000 or less, so very pure materials are necessary to begin with. The impurities have the property of giving up an electron with a far smaller expenditure of energy than the band gap, so that at room temperature most of the electrons are free. A semiconductor of this kind is called 'extrinsic'. It works in the following way. Silicon and germanium, like carbon, have four electrons outside a closed shell. All four elements crystallize in the diamond structure (fig. 53) in which each atom has four nearest neighbours. Homopolar bonds, each making use of two electrons, one from each atom, are formed between

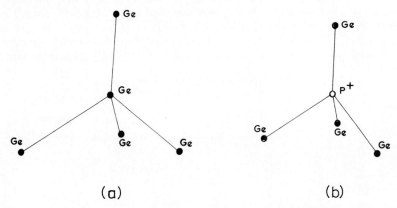

(a)   (b)

Fig. 53. (a) Arrangement of atoms in a germanium crystal (which has the same structure as diamond). (b) Position of phosphorus atom (marked P) in doped germanium.

each pair of atoms. To produce an extrinsic semiconductor one must add an element with a different valency, such as phosphorus which has five electrons outside a closed shell. When phosphorus is added to molten germanium and the material crystallizes, the phosphorus atoms take up sites in the crystal which would otherwise be occupied by a germanium atom, as in fig. 53 (b). Of the five outer electrons in the phosphorus atom, four would be taken up in bond formation; we have to ask what happens to the fifth. If it were separated from the phosphorus atom, it would have an energy in the 'conduction band' of energies (fig. 52), and a wave function of the kind illustrated in fig. 49. It would be free to move about in the crystal, and take part in an electric current. But the phosphorus ion P^+ has a positive charge, and the electron is attracted by this charge. Its potential energy in the field of the P^+ ion is not $-e^2/4\pi\varepsilon_0 r$, but $-e^2/4\pi\varepsilon_0\varepsilon_r r$, where ε_r is the relative permittivity (dielectric constant) of germanium. This is quite high ($\varepsilon_r = 17$) and the semiconducting materials used for transistors must be chosen among those for which the relative permittivity is high. It is then argued that the electron can exist in a series of stationary states in the field of the phosphorus atom, almost exactly like those of an electron in the hydrogen atom in the field of the proton, except that everywhere in the formulae ε_0 has to be replaced by $\varepsilon = \varepsilon_r\varepsilon_0$. Thus the Bohr radius, which gives the radial extent of the wave function, is $\varepsilon_r a_0$, and the binding energy is W_H/ε_r^2, where W_H is ionization energy of hydrogen and a_0 the hydrogen radius. Actually, both theory and experiment have shown that an electron with its energy in the 'conduction band' of germanium or silicon behaves as if its mass were about 0·1 of the normal electronic mass. Thus the Bohr radius becomes

$$0\cdot53 \times 17/0\cdot1 = 90 \text{ Å}$$

and the binding energy

$$13\cdot4 \times 0\cdot1/17^2 = 0\cdot0046 \text{ eV}.$$

The latter is smaller than kT (0·025 eV) at room temperature, so that nearly all the electrons will be free. But at liquid helium temperatures most will be still bound to the phosphorus atoms.

10.4 *An unsolved problem*

It follows from what has been said that germanium or silicon containing phosphorus ought to behave like a solid of swollen hydrogen atoms, not of course arranged as atoms are in a crystal, but distributed much more at random like the atoms in a gas. If so, the considerations of this chapter suggest that a band of energy levels ought to be formed, and that the band, with *one* electron per centre, will be half full. Therefore the electrons ought to be free to move, even at the very lowest temperatures. Actually, if the concentration is high enough so that the distance between the phosphorus atoms is not much greater than the calculated Bohr radius (90 Å), this is so; the resistivity of germanium containing more than 10^{23} phosphorus atoms per m^3 remains quite low, about 10 Ω m. But when experiments are made on specimens with lower concentration of phosphorus, at a certain concentration near 10^{23} m^{-3} the usual semiconducting behaviour appears; the resistivity rises at low temperatures and looks as if it would increase without limit.

This behaviour, together with similar phenomena, is not at the time of writing completely understood. One can see that in a metal there is always a chance of two or more electrons being in one atom at the same time; electrons repel each other, so this will make a positive contribution to the energy. If the electrons are localized, one on each centre, this term in the energy will be absent; on the other hand, the localization within a space of diameter a introduces kinetic energy, a multiple of $h^2/8m_e a^2$. So for some value of a, for which these two quantities are comparable, the transition from metallic to non-metallic should occur. Actually it is observed to occur when the 'expanded Bohr radius' $\varepsilon_r a_0$ is about a quarter of the mean distance between the impurities both for silicon and germanium. But a really complete description of this so called 'metal–insulator transition' awaits further researches.†

† See, for instance, N. F. Mott and Z. Zinamon, *Reports on Progress in Physics* **33**, 881, 1970.

quantum mechanics and the uncertainty principle

THE discussion in Chapter 5 has shown how the paths of a beam of particles can be calculated in certain simple cases. But obviously quantum mechanics cannot limit itself to the properties of beams; like Newtonian mechanics it is a system of mechanics which can be applied to any problem about the motion of particles. Now a typical problem in Newtonian mechanics is that illustrated in fig. 54. A projectile is fired with velocity v at an angle θ to the horizontal; the problem is, what will its range be, and what will be its time of flight?

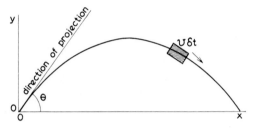

Fig. 54. Trajectory of a projectile fired at an angle θ to the horizontal.

The first of these questions *can* be answered in terms of the theory of beams of particles. We can imagine a machine gun firing a stream of bullets, each leaving the muzzle with velocity v. Each bullet has total energy $\frac{1}{2}mv^2$, and the potential energy gained at a point (x, y, z) is

$$V(x, y, z) = mgy,$$

where y is the height above the gun and g the acceleration due to gravity. The path of the beam can be calculated as in Chapter 5, and the path will be the same as in Newtonian mechanics.

But to describe the time of flight in terms of the behaviour of a wave we shall have to extend our analysis. Suppose the gun is fired at 12 o'clock, but timing errors introduce an uncertainty of $\delta t = 0\cdot1$ s in our knowledge of this time. Then we can think of a volume, shaded in fig. 54, of length $v\delta t$ and moving with velocity v; this is the volume in which we know the projectile to be. According to the ideas of quantum mechanics, we ought to suppose that our knowledge of the movement of the projectile can be deduced from the movement of a 'wave pulse', that is to say the movement of a train of waves of limited

107

length. The mathematical description of such a wave pulse will be exactly similar to that of a flash of light, a radar pulse or a sound wave emitted for a limited time. The amplitude of a wave of the kind envisaged is shown in fig. 55.

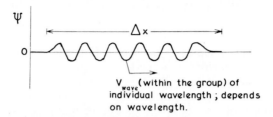

Fig. 55. A wave group of length Δx, illustrating the difference between wave and group velocities.

If quantum mechanics is to make sense, the velocity with which the group moves has to be the same as the observable velocity v of the particles. We have not as yet discussed the frequency of these waves; we must define the frequency in such a way as to obtain this result. But first of all, we have to emphasize that for wave motion in general the velocity with which each wave crest moves (the wave velocity, $v_{\text{wave}} = \lambda v$) is not necessarily the same as velocity of the group as a whole (the group velocity, v_{group}); the two will differ if there is any variation of the wave velocity with wavelength. A plane wave can be written in the usual form, where A is the amplitude and K the wave number,

$$\psi = A \sin \{2\pi(Kx - vt)\}.$$

Then two waves of equal amplitudes A and slightly different frequencies v, v' will, when superimposed, give a wave function equal to

$$\psi = A[\sin \{2\pi(Kx - vt)\} + \sin \{2\pi(K'x - v't)\}], \qquad (1)$$

where K, K' are the wave numbers. (1) may be written

$$\psi = 2A \sin 2\pi \left(\frac{K+K'}{2} x - \frac{v+v'}{2} t\right) \cos 2\pi \left(\frac{K-K'}{2} x - \frac{v-v'}{2} t\right). \qquad (2)$$

This has the form illustrated in fig. 56; the wave described by (2) consists of a train of pulses, the length Δx of each pulse being given by

$$\Delta x = 1/(K-K') \qquad (3)$$

and the velocity of v_{group} of each pulse by

$$v_{\text{group}} = (v - v')/(K - K')$$

108

If ν is very nearly equal to ν', we have therefore approximately

$$v_{\text{group}} = d\nu/dK. \tag{4}$$

A single pulse moves with just the same velocity as each of a succession of pulses. The proof of this will not be attempted here, but is given in numerous textbooks.

The frequency is related to the wave number K by the equation

$$\nu = v_{\text{wave}}K.$$

Thus if the wave velocity is independent of wave number, as is the case for instance for light in a vacuum, the group and wave velocities are the same (for light, 3×10^8 m/s). But if the wave is in a medium which shows dispersion, so that v_{wave} depends on the frequency and hence on K, they are not the same; each wave crest may move faster or slower than the wave pulse as a whole.

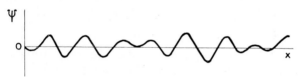

Fig. 56.   A succession of wave groups.

For electron waves, apart from de Broglie's conjecture mentioned in Chapter 4, we know *a priori* nothing about ν. But we do know that, if wave mechanics is to make sense, the group velocity *must* be the same as the particle velocity v; thus

$$\frac{d\nu}{dK} = v.$$

We know, too, from experiment that $\lambda = h/mv$, so that

$$v = hK/m.$$

Therefore

$$\frac{d\nu}{dK} = \frac{hK}{m}.$$

This equation can be integrated; we find

$$\nu = \tfrac{1}{2}hK^2/m + \text{const.}$$

Substituting for K and multiplying by h,

$$h\nu = \tfrac{1}{2}mv^2 + \text{const.} \tag{5}$$

Apart from the constant of integration, therefore, the frequency ν of an electron wave, multiplied by Planck's constant, can be equated to the kinetic energy of each of the particles which the wave describes.

We have now to consider the value which must be given to the constant of integration. It is a constant at a given point of space, but may vary with position.

If a beam of particles is passing through a region of space where there is an electric field of any kind, the kinetic energy changes from point to point. But the frequency of a steady beam cannot change from point to point; in the situation described in fig. 15 of Chapter 4, for example, it is the same at one side of the grid as at the other. The frequency of the wave, therefore, ought to correspond to some property of the particle which remains constant as the particle moves through the electric field. The total energy W of each particle, given by

$$W = \tfrac{1}{2}mv^2 + V(x, y, z),$$

is such a quantity. It is therefore reasonable, in formulating wave mechanics, to choose the constant in (5) as equal to the potential energy $V(x, y, z)$ at the point in space under consideration. We are thus led to the conclusion that, for the frequency v of a de Broglie wave,

$$hv = W. \tag{6}$$

Now that we have a formula for the frequency, we can begin to formulate the framework of quantum mechanics which will replace Newtonian mechanics. Newtonian mechanics starts by considering a particle, at a given point in space and with a known velocity. Newton's laws of motion then make it possible to calculate the trajectory of the particle as in fig. 54. Quantum mechanics proceeds differently. It envisages a *measurement* which determines where the particle is, not exactly but with an uncertainty Δx in its position. It also envisages a measurement of the momentum $p(= mv)$ of the particle. The result of these measurements must, as usual, be described by a wave, the wave group of fig. 55, the wavelength being given by de Broglie's relationship as h/p. Then the laws of wave motion, as we have enunciated them, should be sufficient to predict where and how fast the wave group moves. Wherever the wave group is, $|\psi|^2$ gives the chance of finding the particle at any point.

This formulation at once brings us face to face with the Uncertainty Principle. The wave group of fig. 55 does not have an exactly defined wavelength. We have seen (equation (2)) that a sequence of wave groups each of length Δx is made by combining two plane waves with wave numbers differing by ΔK, where

$$\Delta x \Delta K \simeq 1. \tag{7}$$

This same relationship is true for a single pulse. It is made up of plane waves with wave numbers spread over a range ΔK given by (7). This is a general property of all waves. For instance, a pulse of light of length Δx is not strictly monochromatic; it has its wave numbers spread over a range Δx, with a corresponding spread in its frequencies.

110

The application to quantum mechanics is as follows. Since the wave number K for the wave and the momentum p of the particle are related by the equation $K = p/h$, (7) becomes

$$\Delta p \, \Delta x \sim h. \tag{8}$$

This means that a wave group, describing the results of measurements which determine the simultaneous position (x) and momentum (p) of a particle, cannot be written down if the product of the errors Δx, Δp in the two measurements is less than the Planck constant h. So we are driven to the following conclusion. If wave mechanics is true, it must be *impossible in principle* to measure the position and the momentum at the same time with complete accuracy, and equation (8) gives the greatest accuracy possible.

This was first recognized by the German physicist Werner Heisenberg and the impossibility of this simultaneous measurement is called the Heisenberg Uncertainty Principle. It was first enunciated in 1927. Heisenberg and other scientists at that time examined all sorts of idealized ways of observing these quantities simultaneously to see if the principle could be broken down. For instance, Heisenberg imagined that one might attempt to 'see' an electron under a microscope by illuminating it with light. The electron might be in a beam which had been accelerated by a known field, so that its momentum p is known. Light is shone on the electron as shown in fig. 57, and, the light that it scatters is brought to a focus on a photographic plate. The accuracy Δx with which one can determine the position of the electron depends on the resolving power of the microscope; it is

$$\Delta x = \lambda_{\text{light}} \, d/R,$$

where R is the diameter of the lens aperture, d the distance between the lens and the object that is being viewed and λ_{light} is the wavelength of the light, which we can write c/v. It follows that

$$\Delta x = cd/vR. \tag{9}$$

To obtain high accuracy one would use as short a wavelength as possible, so this idealized experiment was called an observation with a 'gamma-ray microscope'.

Now an observation of the position of an electron cannot be mad without disturbing the electron, because the light quantum itself has momentum hv/c (Chapter 2), and when it is scattered by the electron some of this momentum is transferred to it. If we could allow for this, it would not matter; we should still know the momentum of the electron after it had been observed. But we cannot allow for it completely, because the light quantum may be moving in any of the directions shown by arrows in fig. 57. There is therefore an uncertainty in the

111

direction in which the quantum is going equal to R/d, and therefore an uncertainty Δp in the momentum transferred to the electron along the x axis given by

$$\Delta p = \frac{h\nu}{c} \cdot \frac{R}{d}.$$

Comparing this equation with (9), we see that

$$\Delta p \, \Delta x \sim h.$$

So this particular idealized experiment is not capable of upsetting the Uncertainty Principle.

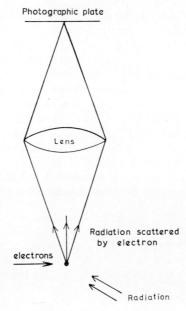

Fig. 57.   The gamma ray microscope.

A book of this kind is not the place to explore in detail the philosophical implications of Heisenberg's Uncertainty Principle. The implication of Newtonian mechanics, that the whole universe including all living matter is an assembly of particles of which the future for all time is determined by the present position and velocity of its parts, is certainly no longer tenable.   This in the 1920's was very upsetting to some of the older generation of scientists, including Einstein, one of the very greatest, and particularly to Schrödinger, who said to Bohr "If we are going to stick to this damned quantum-jumping, then I regret that I ever had anything to do with quantum theory." Heisenberg's view of the matter† is summed up as follows—the Copenhagen interpretation being that which he and Niels Bohr had worked out together and of which the outlines have been sketched in this chapter:

"The criticism of the Copenhagen interpretation of the quantum theory rests quite generally on the anxiety that, with this interpretation, the concept of 'objective reality' which forms the basis of classical physics might be driven out of physics. As we have here exhaustively shown, this anxiety is groundless, since the 'actual' plays the same decisive part in quantum theory as it does in classical physics. The Copenhagen interpretation is indeed based upon the existence of processes which can be simply described in terms of space and time, i.e. in terms of classical concepts, and which thus compose our 'reality' in the proper sense. If we attempt to penetrate behind this reality into the details of atomic events, the contours of this 'objectively real' world dissolve—not in the mist of a new and yet unclear idea of reality, but in the transparent clarity of a mathematics whose laws govern the possible and not the actual. It is of course not by chance that 'objective reality' is limited to the realm of what Man can describe simply in terms of space and time. At this point we realize the simple fact that natural science is not Nature itself but a part of the relation between Man and Nature, and therefore is dependent on Man. The idealistic argument that certain ideas are *a priori* ideas, i.e. in particular come before all natural science, is here correct. The ontology of materialism rested upon the illusion that the kind of existence, the direct 'actuality' of the world around us, can be extrapolated into the atomic range. This extrapolation, however, is impossible."

One further point of great importance arises out of the analysis of this chapter. An approximate measurement of the position and the momentum of a particle make it possible to write down the form of the wave pulse shown in fig. 55; the laws of wave motion should then predict how the wave group moves. But it is not enough just to know the initial form of the wave; the wave shown in fig. 55 might just as well be moving from left to right or from right to left. In order to predict the future behaviour of the wave, we need the *two* functions f and g introduced in Chapter 5 as the real and complex forms of the wave function, analogous to the displacement y and velocity $\dot{y}$ of any point on a string. In order that the laws of Newtonian mechanics should predict the future behaviour of a wave on a string, both y and $\dot{y}$ must be known initially; in the same way f and g must be known for the future behaviour of the wave function to be determined.

We turn now to the wave equation which, we have asserted, predicts the future form of the wave function if the complex wave function

† From *Niels Bohr and the Structure of the Atom*.

$\Psi(=f+ig)$ is given initially.† We can find it as follows. For a wave representing particles all of the same energy W, Ψ is of the form

$$\Psi = \psi(x, y, z) \exp(-2\pi i \nu t) \tag{10}$$

with $\nu = W/h$. The quantity ψ satisfies

$$\nabla^2\psi + \frac{8\pi^2 m}{h^2}(W - V)\psi = 0. \tag{11}$$

The equation we require must not contain W, since it does not describe particles for which the energy is known exactly, and must contain $\partial\Psi/\partial t$, since it shows how Ψ varies with time. So differentiating (10) we have

$$\frac{\partial\Psi}{\partial t} = -\frac{2\pi i W}{h}\Psi$$

and using this equation to eliminate W from (11) we find

$$\frac{h}{2\pi i}\frac{\partial\Psi}{\partial t} = H\Psi, \tag{12}$$

where H is written for

$$H = -\frac{h^2}{8\pi^2 m}\nabla^2 + V. \tag{13}$$

Equation (12) is the 'time dependent' Schrödinger equation. The most important thing about it is that it is of the first order in $\partial/\partial t$. This means that, if Ψ at all points is known at one moment of time, the equation gives the rate of change of Ψ at all points, and so it gives the value of Ψ everywhere at a small interval of time later. Thus it does just what we want, namely to predict the *movement* of the shaded area of fig. 54, that is the wave group of fig. 55, which represents the parts of space where the particle is likely to be.

† A capital Ψ is used to distinguish a wave function that changes with t from one (ψ) that does not.

114

absorption of radiation

EQUATION (12) of the last chapter has to be used to describe any situation in which the probability that a system is in a given state changes with time. There are many physical problems where this is so, and we shall discuss here only one, the absorption of radiation by the atoms of a gas. The Einstein B coefficient, which defines the intensity of absorption, has been described in Chapter 2; we have now to show how to calculate it.

We suppose that an atom is initially in its ground state, and that the wave function of an electron in this atom is

$$\Psi_0 = \psi_0(r) \exp\left(-2\pi i W_0 t/h\right). \tag{1}$$

It is acted on by a light wave. For the purpose of this discussion, we consider only the electric vector in the light wave which we denote by E. The wavelength of visible light (say 4×10^{-7} m), is much smaller than the radius of the atom (10^{-10} m). Therefore it is a good approximation to suppose that, at a given instant of time, the electric field E is the same at all points within the atom, and if it is along the x axis, the potential energy of the electron within the atom due to the field E can be written

$$V(x) = eEx.$$

But E is varying periodically with time with the frequency ν of the light wave. We therefore write

$$E = E_0 \cos 2\pi\nu t$$

and

$$V(x) = eE_0 x \cos 2\pi\nu t. \tag{2}$$

The time-dependent Schrödinger equation then becomes (see equation (13) on p. 114)

$$\frac{h}{2\pi i} \frac{\partial \psi}{\partial t} = H\psi + eE_0 x \cos 2\pi\nu t. \tag{3}$$

Under the influence of the final term due to the field the form of the wave function will gradually change; it will no longer have the form (1), representing an atom in the ground state, but will take up the more general form

$$\Psi = \sum_n A_n(t)\psi_n(x, y, z) \exp\left(-2\pi i W_n t/h\right). \tag{4}$$

115

This represents a state of the electron in which *all* the normal modes of the wave are excited, that designated by n having amplitude $A_n(t)$; it is in fact the most general solution of the Schrödinger equation (3).

The interpretation of such a solution is as follows. $|A_n(t)|^2$ has to be taken as the probability, at time t, that the atom is in the state n. Within the context of this book, this must be seen as a new assumption, but it points towards the need for a more general formulation of quantum mechanics where a wave function can be used to determine the probability that *any* observable property of the system, such as position, angular momentum, energy or anything else, lies within a range of values. Such a formulation is given in other text books.

It remains to be shown how the coefficients can be calculated and the B coefficient deduced. First of all, the field of the light wave is very small compared with the fields within the atom, so the perturbing terms (2) can be treated as small, and at any rate initially, when the atom has not been exposed to the radiation for too long, the wave function (4) will not differ much from (1). So, inserting (4) into (3), and neglecting very small quantities, we find

$$\left(\frac{h}{2\pi i} \cdot \frac{\partial}{\partial t} - H\right)\Psi = V(x)\Psi_0. \qquad (5)$$

Substituting for Ψ and remembering that

$$\left(\frac{h}{2\pi i} \cdot \frac{\partial}{\partial t} - H\right)\Psi_n = 0,$$

(5) reduces to

$$\frac{h}{2\pi i} \sum_n \frac{dA_n}{dt} \, \psi_n \exp\left(-2\pi i W_n t/h\right). \qquad (6)$$

We cannot proceed any further without using the important 'orthogonal' property of the wave function, namely that if ψ_n, $\psi_{n'}$ are two non-degenerate real solutions of the Schrödinger equation, then

$$\iiint \psi_n \psi_{n'} \, dx \, dy \, dz = 0, \qquad (7)$$

where the integral is over all space. This can be proved very simply in one-dimension. The Schrödinger equations which the two solutions satisfy are then

$$\frac{d^2\psi_n}{dx^2} + \frac{8\pi^2 m}{h^2}\{W_n - V(x)\}\psi_n = 0,$$

$$\frac{d^2\psi_{n'}}{dx^2} + \frac{8\pi^2 m}{h^2}\{W_{n'} - V(x)\}\psi_{n'} = 0.$$

116

Multiplying the first equation by $\psi_{n'}$ and the second by ψ_n, and subtracting, we find

$$\left\{\psi_{n'}\frac{d^2\psi_n}{dx^2} - \psi_n\frac{d^2\psi_{n'}}{dx^2}\right\} + \frac{8\pi^2 m}{h^2}(W_n - W_{n'})\psi_n\psi_{n'} = 0. \tag{8}$$

The term in the curly brackets is equal to

$$\frac{d}{dx}\left\{\psi_{n'}\frac{d\psi_n}{dx} - \psi_n\frac{d\psi_{n'}}{dx}\right\}. \tag{9}$$

This shows that if we integrate (9) over the range in which x is defined, we obtain zero, since ψ_n, $\psi_{n'}$ tend exponentially to zero as x or $-x$ becomes large; or if the problem is that of a particle shut up in box, ψ_n will vanish at its boundaries. We have supposed that the states are non-degenerate, which means that $W_n \neq W_{n'}$. On integrating (8), the orthogonal relation (7) follows.

Making use of (7) and assuming it to be true in three dimensions, we now multiply (6) by $\psi_n \exp(2\pi i W_n t/h)$ and integrate over all space. We find

$$\frac{h}{2\pi i} \cdot \frac{dA_n}{dt} = eE\langle n|x|0\rangle \cos 2\pi\nu t, \exp\left\{-\frac{2\pi i(W_n - W_0)t}{h}\right\} \tag{10}$$

where $\langle n|x|0\rangle$ is written for the quantity given by

$$\langle n|x|0\rangle = \int \psi_n x\psi_0 \, dx \, dy \, dz. \tag{11}$$

The quantity (11) is called the matrix element of x. From equation (10) several results follow:

(i) A_n will only increase with the time if

$$\nu = \pm(W_n - W_0)/h, \tag{12}$$

because otherwise all terms on the right-hand side of equation (10) oscillate with time. This means that our equation only predicts transitions if this equation (12) is satisfied. The equation

$$h\nu = W_n - W_0$$

is simply the Bohr frequency condition for the absorption of light. The equation

$$h\nu = W_0 - W_n$$

applies if the initial state, with energy W_0, lies *above* the final state, and shows that radiation can stimulate an atom to jump downwards, with the emission of a quantum.

117

(ii) Only for certain combinations of states does the integral (12) have finite values. For instance, if the states are both s-states, ψ_n and ψ_0 are both spherically symmetrical and the integrand is positive for $x > 0$, negative for $x < 0$ and the integral vanishes. Similarly if both states are p-states the integral vanishes too. s–s and p–p transitions are said to be forbidden. On the other hand s–p transitions are allowed, since if ψ_0 is spherically symmetrical and ψ_n of the form $xf(r)$, then

$$\int \psi_n x \psi_0 \, dx \, dy \, dz$$

does not vanish.

The rules which exclude optical transitions between s-states or between p-states, but which allow transitions between s and p-states are called 'selection rules'. Their general formulation, and the solution of (10) to give the B coefficient, is explained in more advanced text-books.

INDEX

THE WYKEHAM SCIENCE SERIES

THE WYKEHAM TECHNOLOGY SERIES

All orders and requests for inspection copies should be sent to the appropriate agents. A list of agents and their territories is given on the verso of the title page of this book.